LASERS
&
HOLOGRAPHY
An Introduction to Coherent Optics

Winston E. Kock

Director of the Herman Schneider Laboratory
University of Cincinnati

(Second, Enlarged Edition)

Dover Publications, Inc.
New York

To Sir C. V. Raman

International Standard Book Number: 0-486-24041-X
Library of Congress Catalog Card Number: 80-69746

Manufactured in the United States of America
Dover Publications, Inc.
180 Varick Street
New York, N.Y. 10014

PREFACE

Holography and photography: two ways of recording, on film, information about a scene we view with our eyes. Yet, how different the basic mechanism by which they accomplish their purpose, and how different the images which result. As the words *holo* (complete) and *gram* (message) connote, the hologram captures the entire message of the scene in all its visual properties, including the realism of three dimensions. The photograph, on the other hand, collapses into one plane, the plane of the print, all the scenic depth we perceive in the actual scene.

Perhaps the first demonstration of photography by the French scientist, Louis Jacques Mandé Daguerre, in 1839, lulled many who followed him into assuming that he had found the only, the ultimate, procedure. Daguerre sought to record, and succeeded in recording, the image formed on the ground glass screen of a much older invention, the *camera obscura*, a device to assist artists in drawing more lifelike pictures of the scenes before them. The *camera obscura* was a box fitted with a lens at one end and a slanted mirror at the other. The image was then "collapsed" onto a planar ground glass screen at the top of the box, where the artist could study it or trace it onto translucent paper. Daguerre's process of photography just recorded, on his new, light-sensitive material, this collapsed image. Many probably thought: "Why ask for more?"

Not until 1947 did the British scientist, Dennis Gabor, conceive of holography, a new and ingenious method for photographically recording a three-dimensional image of a scene. As is true of most exceptional ideas, it is hard to understand why holography had not been thought of sooner. As we shall see, it involves merely the quite simple process of photographically recording the pattern formed by two interfering sets of light waves, one of these wave sets being a *reference wave*. Now, a half-century before Gabor, the French scientist, Gabriel Lippmann, had proposed recording photographically a light-wave interference pattern (a standing wave pattern) for a form of color photography. And several decades before holography, the technology of radio had made extensive use of reference waves. So the elements of holography were all there, waiting for an ingenious mind to assemble them into a brilliant idea.

All young scientists and engineers should ponder this point well. For however holography may benefit mankind in the future, it has already proved that opportunities continually exist for combining old concepts into new, valuable ideas. It has shown that one need not have full command of the very forefronts of science to contribute significantly to progress, that even a modest understanding of relatively simple processes permits the *ingenious* person to contribute in today's expanding world of technology.

Although Gabor conceived his idea rather recently, he was still too early. For the special kind of light needed to demonstrate the full capabilities of holography, a single-frequency form called *coherent* light, was not available in abundance in 1947. It became available only after the *laser*, a new light source first demonstrated in 1960, was developed. An atomic process called *stimulated emission* is responsible for the light generation in a laser, and the name *laser* is an acronym formed from the words *L*ight *A*mplification by the *S*timulated *E*mission of *R*adiation. Because the light sources available to Gabor in 1947 could not fully demonstrate holography, it lay almost dormant for many years. In 1963, the American scientist, Emmett Leith, introduced the laser to holography. The subsequent advances made by him, by another American scientist, George Stroke, and by their many co-workers, led to a tremendous explosion in holography development, having its "ground-zero" at the original Leith-

Stroke base of operations, the University of Michigan in Ann Arbor.

A hologram records the interference pattern formed by the combination of the reference wave with the light waves issuing from a scene, and when this photographic record is developed and again illuminated with laser light, the original scene is presented to the viewer as a *reconstructed image*. This image manifests such vivid realism that the viewer is tempted to reach out and try to touch the objects of the scene. The hologram plate itself resembles a window with the imaged scene appearing behind it in full depth. The viewer has available to him many views of the scene; to see around an object in the foreground, he simply raises his head or moves it left or right, in contrast to the older, two-photograph stereo pictures which provide an excellent three-dimensional view of the scene, but only *one* view. Also, photography uses lenses, and lenses allow only objects at a certain distance from the camera to be in truly sharp focus. In the hologram process, no lenses are used, and all objects, near and far, are portrayed in its image in extremely sharp focus.

Gabor first used the term *hologram,* and although various words could have been devised to describe the general hologram process—for example, hologrammetry—hologram pioneer George Stroke proposed *holography,* and this has become the generally accepted term. To give the reader a general overview of the hologram process, the first chapter will describe holography in two of its simplest forms. Later chapters will review some of its underlying wave concepts, including coherence, diffraction, interference, and the way lasers generate their special coherent light. The nature of holograms is discussed in Chapter 6, and a concluding chapter brings us up to date on developments in the field.

Acknowledgments

This book is an outgrowth of my presentation at the U.S.-Japanese seminar on holography held in Tokyo in October 1967 under the auspices of the U.S. National Science Foundation and the Japanese Society for the Promotion of Science. I wish to thank Professor George Stroke, State University of New York at Stony Brook and head of the U.S. delegation, for the invitation to participate. I am also indebted to Mr. Floyd K. Harvey and the Bell Telephone Laboratories for the use of the photographs portraying sound waves and micro-waves; Messrs. Lowell Rosen and John Rendeiro of the NASA Electronics Research Center, and Mr. Stan Krulikowski of the Bendix Research Laboratories for various hologram photographs; and the National Aeronautics and Space Admin-istration for *Plates* 14, 27, and 32.

CONTENTS

The classical zone plate. Zone plates as negative lenses. Zone plates with areas interchanged. Offset zone plates. Offset holograms. Zone plates as lenses. Volume zone plates. Standing wave patterns. Lattice reflectors. Reflection zone plates.

Three-dimensional realism. Holograms and photographs. Parallax and lens action. A stereo hologram. Focused-image holography. Reconstruction with a small portion of a hologram. Pseudoscopy in the real image of a hologram. Image inversion. Single wavelength nature of holograms. Requirements of film properties. Information content. Holograms and coherent radar. Acoustic holograms.

Viewing holograms. The Museum of Holography. White light or "rainbow" holograms. Hologram computer memories. Liquid surface holography. Synthetic-aperture radar and sonar. New lasers. Wires of glass. Lasers in the military.

Chapter 1

HOLOGRAMS AS WAVE PATTERNS

In this chapter, holography will be described in very simple terms; a more thorough treatment of its many unusual aspects will be presented in succeeding chapters. Here we shall consider holography simply as the photographic recording of an *interference pattern* between two sets of light waves.

Properties of Waves

Certain basic properties of wave motion are manifested by water waves created when a pebble is dropped on the surface of a still pond, as shown in Figure 1. Because all wave

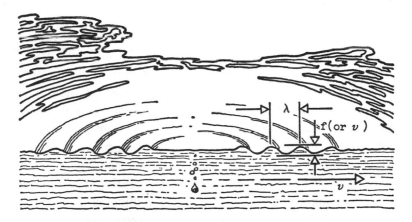

Figure 1. Water waves on a pond. The speed v is called the velocity of propagation; the distance from crest to crest, the wavelength λ; and the periodicity of the up-and-down motion of a point on the surface, the frequency f (or v).

energy travels with a certain speed, such water waves move outward with a wave speed or *velocity of propagation, v.* Waves also have a *wavelength,* the distance from crest to crest; it is usually designated, as shown, by the Greek letter λ. If, in Figure 1, we were to position one finger so that it just touched the crests of the waves, we could feel each crest as it passed by. If the successive crests are widely separated, they touch our finger less frequently than if the crests are close together. The expression *frequency (f* or the Greek letter *nu, ν)* is therefore used to designate how often (how many times in one second) the crests pass a given point.

Obviously, the velocity, the wavelength, and the frequency (stated as *cycles per second*) are related by the expression:

frequency equals velocity divided by wavelength   (1).

This says simply that the shorter the wavelength, the more frequently the wave crests pass a given point, and similarly, the higher the velocity, the more frequently the crests pass. No proportionality constant is needed in equation (1) if the same unit of length is used for both the wavelength and the velocity and if the same unit of time (usually the second) is used for both the frequency and the velocity.

Water waves move fairly slowly; their progress readily can be observed on a still body of water. Sound waves travel much faster. Their speed, the speed of sound, is eleven hundred feet per second (over six hundred miles an hour). A sound wave having a frequency of 1100 cycles per second (this is approximately the frequency of a note two octaves above middle C on the piano) has a wavelength of one foot. Light waves have the highest velocity of all, one hundred and eighty-six thousand miles a second. Violet light has the extremely short wavelength of sixteen millionths of an inch. Because a hologram is a photographic recording of a light wave pattern, we shall see that the extremely short wavelengths of visible light place rather severe requirements on the photographic plate used in the process.

Wave Uniformity

The water waves of Figure 1 are somewhat unusual. First, they are extremely regular; second, they are waves of only one wavelength. Had there been strong winds blowing on the pond, waves of many different wavelengths would have been observed, and they would, in general, have been traveling in many different directions, causing irregular wave patterns. For other types of wave motion, such as sound waves or light waves, such uniform, single-wavelength waves are also rather unusual. The sound waves of noise, for example, are multiwavelength and very irregular, as are the light waves issuing from ordinary incandescent lamps.

Nevertheless, single-wavelength sound waves and light waves *can* be generated, and they then behave exactly like the water waves of Figure 1. The space pattern of uniform, single-frequency sound waves, radiating from the receiver unit of a telephone headset, is shown in *Plate* 1. This sound wave photo was made by minutely scanning the entire area of the sound field with a sound-sensitive microphone, converting the microphone signal to light intensity, and photographically recording the varying light signal.* The resemblance of this pattern to the pattern of water waves in Figure 1 is evident; the white areas can be considered as crests of the waves and the dark areas as troughs. A similar circular pattern of single-wavelength microwaves is shown in *Plate* 2; these microwaves are very short wavelength radio waves, and, like light waves, are *electromagnetic* in nature (rather than mechanical as are the sound waves of *Plate* 1).

Interference

Now, if the simple and uniform set of sound waves of *Plate* 1 or the set of electromagnetic waves of *Plate* 2 were to meet a

* For further details on this wave-recording technique, see the author's Science Study Series book, SOUND WAVES AND LIGHT WAVES, Doubleday, 1965.

second set of similarly uniform, single wavelength waves, a phenomenon called *interference* would result. At certain points, the two sets would add a condition called *constructive* interference, and at others, they would subtract a condition called *destructive* interference. As sketched at the left of Figure 2, when the crests of one wave set, A, coincide with the crests of a second set, B, constructive interference occurs, and the height of the combined crests increases. When, on

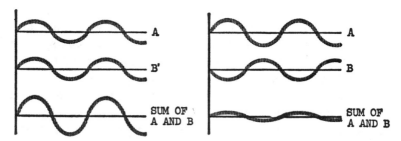

Figure 2. Two waves of the same wavelength add (at the left) if their crests and troughs coincide, and subtract (at the right) if the crest of one coincides with the trough of the other.

the other hand, the crests of one source coincide with the *troughs* of the second source, as shown on the right, destructive interference occurs, and the combined crest height is lowered. For sound waves, such additive and subtractive effects cause increases and decreases in loudness in the sound pattern; for light waves, they cause variations in brightness or light intensity.

When each of the two interfering wave sets is simple, the *interference pattern* (the positions where wave addition and wave subtraction occur) can be predicted and portrayed easily. On the other hand, when the wave sets are complicated, the interference pattern becomes very confused. *Plate* 3 portrays a moderately complicated interference pattern formed by the combination of a simple set of sound waves and the more complicated set of sound waves passing into the shadow area of a wooden disk. (We shall discuss this pattern in more detail in Chapter 4).

Making a Hologram

In forming a *hologram,* two sets of single-wavelength light waves are made to interfere. One set is that issuing from the scene to be photographically recorded; almost invariably, it is an extremely complicated wave set. The other is usually rather simple, often being a set of plane waves. This second set is called the *reference* wave, and, in reproducing or *reconstructing* for the viewer the originally recorded scene, a set very similar to the original reference wave set is used to illuminate the developed photographic plate, the hologram.

The two sets of hologram waves are caused to interfere at the photographic plate, as shown in Figure 3. Here the "scene" comprises a pyramid and a sphere. The objects are

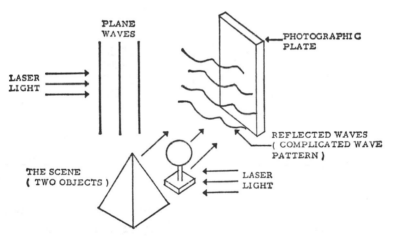

PLANE WAVES

PHOTOGRAPHIC PLATE

LASER LIGHT

REFLECTED WAVES (COMPLICATED WAVE PATTERN)

THE SCENE (TWO OBJECTS)

LASER LIGHT

Figure 3. In making a hologram, the scene is illuminated with laser light, and the reflected light is recorded, along with a reference wave from the same laser, on a photographic plate. The plate is then developed and fixed.

illuminated by the same source of single-wavelength laser light which is forming the plane waves at the top of the figure. Because the wavefronts of the set of waves issuing from the scene are quite irregular, the interference pattern in this case is quite complicated, much more complicated than the pattern of *Plate* 3. After exposure, the photographic plate is developed and fixed, and it thereby becomes the hologram. When it is illuminated, as shown in Figure 4, with the same laser

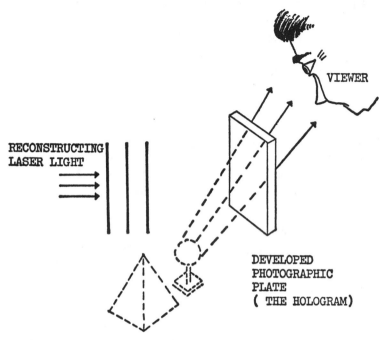

Figure 4. When the developed plate of Figure 3 is illuminated with the same laser reference beam, a viewer sees the original scene "reconstructed," standing out in space, with extreme realism, behind the hologram "window."

light used earlier as the reference wave, a viewer imagines he sees the original two objects of Figure 3 in full three dimensions.

A Photographic Grating

To understand how such a light wave interference pattern, once photographically recorded and then developed, can later re-create a lifelike image of the original scene, we must first examine two rather simple interference patterns. The first is that formed by combining two sets of plane waves. As sketched in Figure 5, the combination of sets A and B causes wave *addition* to occur along those horizontal lines of the photographic plate where the positive crests of the two sets reinforce each other (marked ++). Wave diminution occurs where the positive crest of one meets the negative trough of the other (marked +−). The light intensity is greater

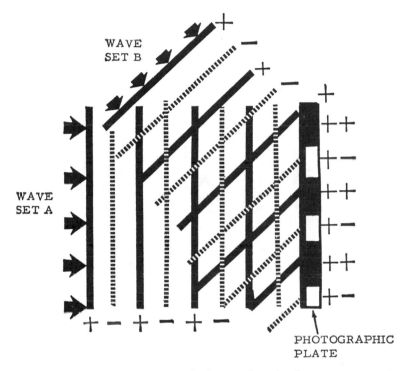

Figure 5. When two sets of single-wavelength plane waves meet, interference occurs; where wave crests and troughs coincide, wave addition results; and where a crest of one coincides with a trough of the other, wave cancellation occurs.

along those lines where the light energy adds, and accordingly, the plate is more strongly exposed there. Conversely, along those lines where a diminution of energy exists, the plate is more weakly exposed. Parallel striations of light are thus recorded on the photographic film, and after the plate is developed and fixed, these striations appear on the film as lines. A photographic record of this type is shown in *Plate* 4.

The photographic record of *Plate* 4 is a form of wave *grating,* a widely used optical device usually made by ruling parallel lines very close together on a piece of glass. If such a ruled grating, or the photographic grating just described, is illuminated by horizontally traveling plane waves, these plane waves are affected by the array of horizontal lines in a well-known way. This is shown in Figure 6. A large portion of

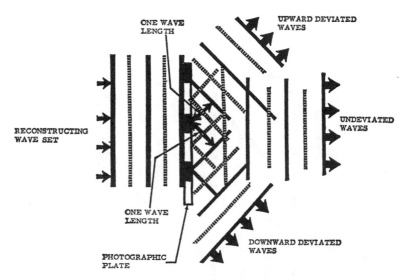

Figure 6. When the line pattern of Figure 4 (as recorded in the process sketched in Figure 5) is illuminated with the original, horizontally traveling set of plane waves, two off-angle plane wave sets and one straight-through plane wave set result. One of the off-angle sets travels in the same direction in which the original off-angle wave set was traveling.

the light wave energy passes straight through the grating, as light would pass through any nonopaque photographic film which has been exposed, developed and fixed. In addition, because the reconstructing waves are again single-wave-length waves, the line pattern will cause some of the wave energy to be deflected (diffracted) at angles *off* the main direction.

We saw in Figure 2 that when wave crests coincide, wave height increases. This same additive condition can also occur if one of the two wave sets *falls behind* the other by a full wavelength (or any integral number of wavelengths). Thus, in the left-hand sketch of Figure 2, addition would still occur even if the lower wave were to be slid one full crest to the left or right. Similarly, for the grating of Figure 6, wave energy passing through the open spaces will add, not only in the forward, horizontal direction, but also for those two directions for which one open area is one full wavelength behind the next higher, or the next lower, open area. In the figure, these two off-axis directions are indicated, one where wave energy is deviated or diffracted upward, and one where it is diffracted downward. For the upward direction, the waves emerging from each transparent area are one wavelength behind those leaving the transparent area immediately *above* it; for the downward direction, waves emerging from each transparent area are one wavelength behind the waves leaving the area immediately *below* it. Waves passing through the transparent areas thus add constructively in the three directions shown, including two sets of diffracted waves.

Note particularly that the downward diffracted set in Figure 6 is proceeding in exactly the same direction as the original set, B, of Figure 5 would have traveled, had the photographic plate not been present. Accordingly, a viewer in the path of these *reconstructed* waves would imagine that the source which generated the original set, B, in Figure 5 was still located behind the hologram. The photographic *hologram grating* is thus able to "regenerate" or reconstruct a wave progression long after it has ceased to exist. The hologram grating of Figure 6 also generates a second set of upward-moving waves which were not present originally.

A Photographic Zone Plate

The second wave interference pattern of importance in holography is that formed by the interference between a set of plane waves and a set of spherical waves. In this case, a circular pattern is formed instead of the parallel line striations of *Plate* 4. A cross section of this circular pattern is sketched in Figure 7. As in Figure 5, parallel plane waves (set A) are again shown arriving from the left, and they interfere at the photographic plate with the spherical waves (set B) issuing from point source, P. Again, as in Figure 5, areas of wave subtraction and addition exist. In this case, however, as the distances increase from the central axis, the separation be-

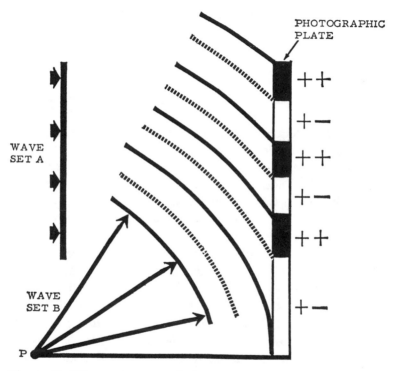

Figure 7. When two sets of single wavelength wave sets, one plane and the other spherical, meet at a plane, a circular interference pattern results, with the separation of the outer circles continually decreasing.

tween these areas lessens. The uniform spacing of striations of *Plate* 4 no longer exists; in addition, the spherical waves generate rings rather than parallel striations.

A photographic recording of an interference pattern between plane and spherical light waves is shown in *Plate* 5; it was made by photographing the pattern generated when plane and point source, single-frequency light waves inter-fere. Whereas the record of *Plate* 4 is called a grating, the record of *Plate* 5 is called a *zone plate*. The similarity be-tween holograms and zone plates was first noted by the British scientist, G. L. Rogers, in 1950. Sections located near the central top and central bottom edges of *Plate* 5 of this pattern resemble somewhat (except for the slight curvature) the horizontal line pattern of Figure 5 and *Plate* 4. Thus, it is to be expected that, as in *Plate* 4, when this pattern is illuminated with plane waves, three wave sets will again be generated.

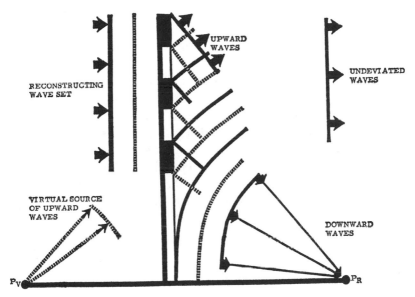

Figure 8. When the upper portion of the circular line pattern of *Plate* 5 (as recorded in the process sketched in Figure 7) is illu-minated with the original, horizontally traveling set of plane waves, three sets of waves result. One set travels straight through horizontally; another acts as though it were diverging from the source point of the original spherical waves; and the third is a set which converges toward a point on the opposite side of the re-corded circular pattern.

This is shown in Figure 8. As before, a portion of the re-constructing plane wave set arriving from the left is undevi-ated, passing straight through the photographic transparency.

The circular striations near the top of the drawing act as do the horizontal lines of *Plate* 4; they cause energy to be diffracted both upward and downward. Because, however, the pattern of striations is circular, the upward waves travel outward as circular wave fronts originating at the point, Pv. These waves form what is called a *virtual image* of the original point light source P (virtual, because in the reconstruction, no source really exists there). These waves give to an observer, located where the words "upward waves" appear in Figure 8, the illusion that an actual point source of light exists there, fixed in space behind the photographic plate no matter how he moves his head. Furthermore, this imagined source exists at exactly the spot occupied by the original spherical wave light source used in making the photographic record.

As occurs with the plane wave interference pattern of Figure 6, a third set of waves is also formed by the spherical wave pattern. In Figure 8 this set is shown moving downward, and because the recorded pattern is circular, these waves are focused waves. They converge at a point which is located at the same distance to the far side of the photographic record as the virtual source is on the near side. The circular striations cause a *real image* of the original light source, P, to form at Pr (*real*—a card placed there would show the presence of a true concentration of light).

The Complete Hologram Process

The complete, two-step hologram process is shown in Figure 9. Here, a pinhole in the opaque card at the left serves as the "scene," a point source of spherical waves. These interfere at the photographic plate with the plane waves arriving from the left. The upper portion of the circular interference pattern is photographically recorded. When the photographic plate is developed and fixed and then placed in the path of plane light waves, as shown in the diagram on the right, a virtual image of the original pinhole light source is formed at the *conjugate* focal point, Fc. A viewer at the

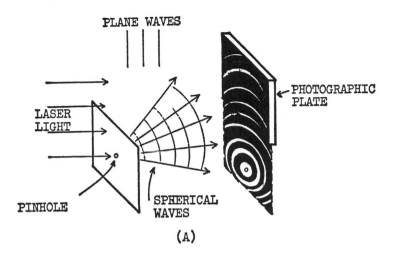

PLANE WAVES

LASER LIGHT

PINHOLE

SPHERICAL WAVES

PHOTOGRAPHIC PLATE

(A)

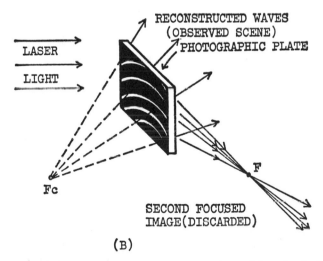

LASER

LIGHT

RECONSTRUCTED WAVES
(OBSERVED SCENE)
PHOTOGRAPHIC PLATE

Fc

F

SECOND FOCUSED
IMAGE (DISCARDED)

(B)

Figure 9. Plane reference waves interfering with spherical waves issuing from a pinhole from a zone-plate interference pattern (A), which, when photographically recorded and reilluminated with laser light, generates waves appearing to emanate from the original pinhole (B).

upper right thus imagines he sees the original light from the pinhole. The real image (the focused image) appears at the true focal point as shown; in the usual viewing of a hologram, this second wave set is not used. In this figure, the straight-through, undeviated waves are not shown.

In Figure 10, a similar photographic recording procedure is sketched, except, for this case, the original scene is one having not a single pinhole, but three pinhole sources of light, each in a different vertical location and each at a different axial distance from the plane of the photographic plate. We

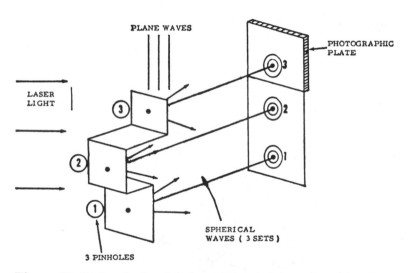

Figure 10. If the single pinhole of Figure 9 is replaced by three separated pinholes, the zone-plate patterns of all three are photographically recorded; when this photograph is re-illuminated, all three pinholes are seen in their correct three-dimensional positions. A more complicated three-dimensional scene can be considered as many point sources of light, each generating, on the hologram plate, its own zone plate; each of these zone plates will then reconstruct its source in its original three-dimensional position.

see that each of the three light sources generates its own circular, many-ring pattern, comparable to the single pattern of Figure 9. (In the figure, only the first two central circular sections of these three patterns are indicated.) The upper portions of the three sets of circular striations (those encompassed by the photographic plate) are recorded. When this film is developed and fixed and then reilluminated, as was done for the single pinhole recording of Figure 9, three sets of upward waves and three sets of focused waves are generated. Of particular importance, from the standpoint of holography, is that virtual images of each of the three pinholes is generated (by the upward, diverging waves), and a real image of each is formed by the downward, converging waves. The virtual images cause a viewer at the top right to imagine that he sees three *actual* point sources of light, all fixed in position and each positioned at a different (three-dimensional) location in space. From a particular viewing angle, source number three might hide source number two. However, if the viewer moves his head sideways or up or down, he can see around source number three and verify that source number two does exist.

The Hologram of a Scene

All points of any scene that we perceive are emitting or reflecting light to a certain degree. Similarly, all points of a scene illuminated with laser light are reflecting light. Each point will have a different degree of brightness; yet, each reflecting point *is* a point source of laser light. If a laser reference wave is also present, each such source can form, on a photographic plate, its own circular interference pattern in conjunction with the reference wave. The superposition of all these circular patterns will form a very complicated interference pattern, but it will be recorded as a hologram on the photographic plate, as was sketched in Figure 3. When this complicated photographic pattern is developed, fixed, and reilluminated, reconstruction will occur and light will be diffracted by the hologram, causing all the original light sources to appear in their original, relative location, thereby providing a fully realistic three-dimensional illusion of the original scene.

Chapter 2
COHERENCE

In the preceding chapter, the light waves discussed in forming holograms were referred to as single wavelength or, what is the same thing, single-frequency waves. Waves which have this single-frequency property are said to have good *frequency coherence*. Thus, light from a laser (particularly a gas laser) is said to exhibit an extremely high degree of frequency coherence. Because of the importance of wave coherence in holography, let us examine what the term coherence means.

Frequency Coherence

We noted that some water waves, some sound waves, and some light waves are single wavelength and rather uniform. One method of indicating the single wavelength nature of a

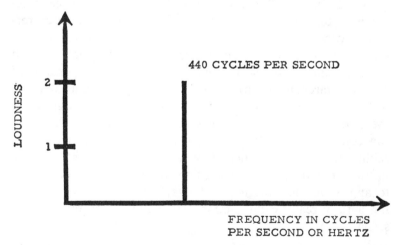

Figure 11. A wave, such as a sound wave, comprising but one single frequency, is represented in this way.

sound wave (the extent of its frequency coherence) is by portraying the *frequency analysis* of the waves. In Figure 11, such an analysis is presented for a single-frequency sound wave. The frequency content is plotted along the horizontal axis, and the amplitude or loudness of each frequency component is specified by the height on the vertical axis. The frequency or pitch of the one-frequency component of this sound is indicated as 440 cycles per second (the note A in the musical scale), and its loudness as two (arbitrary) units. Such a tone would be classed as having an extremely high frequency coherence.

An analysis of a somewhat more complicated sound is shown in Figure 12. This sound is one comprising a funda-

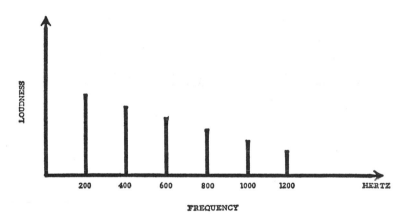

Figure 12. A periodic wave may have many harmonics, all of which are integral multiples of the fundamental frequency.

mental tone (the lowest frequency tone) and many overtones, with all overtones being harmonically related to the fundamental. Like the single-frequency sound, this tone also would be musical in sound because its waves constantly repeat themselves at the rate of the fundamental.

A still more complicated sound is the sound of noise. Noiselike sounds are very irregular, and therefore, they have little coherence. They include, for example, the sound of a jet aircraft or the sound of howling wind in a storm; such sounds have a broad, continuous spectrum of frequencies. The analysis of an exemplary noiselike sound is sketched in Figure 13.

The sound portrayed in Figure 11 is that of a perfectly pure, single-frequency sound; in other words, it is one of *infinitely* high frequency coherence. Actually, such absolutely perfect waves do not exist; however, the extent to which a wave *approaches* this perfection can be specified. Figure 14

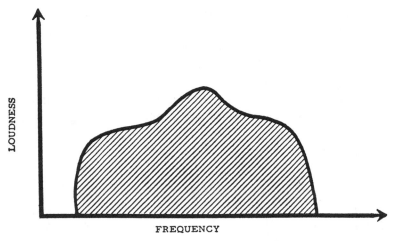

Figure 13. A non-periodic wave can be "noiselike" and possess a broad spectrum of frequencies.

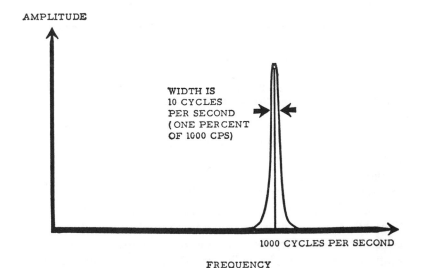

Figure 14. When a non-periodic wave approaches in its nature a periodic wave, its spectrum narrows.

shows the representation of a noiselike sound whose components extend over a fairly narrow frequency band. The extent to which such a signal approaches one having an infinitely narrow band, that is, a single-frequency content, can be specified by stating the width of its frequency band as a percentage of the center frequency. In Figure 14, the bandwidth is ten cycles per second, and the central frequency component is 1000 cycles per second. Thus, the sound is one having a one percent frequency spread or a bandwidth of one percent.

Frequency Stability

Another way of specifying how closely a coherent wave approaches perfection is by stating its frequency constancy or frequency stability. The very fact that there is a frequency spread in the analysis of Figure 14 indicates that the frequency of the tone involved wavers back and forth in pitch over the frequency band shown. It is, accordingly, not absolutely constant in pitch, and the specifying of a one percent bandwidth is equivalent to the statement that the tone has a frequency instability (or stability) of one part in one hundred.

Electronic sound wave generators (called *oscillators*), such as those used in electronic organs, employ various procedures to achieve a high degree of pitch or frequency stability. Such procedures avoid the need for periodic tuning of the instrument. Some audio oscillators achieve a frequency stability of one part in a million or better.

Radio waves, also, usually are generated electronically, and they, too, can be made very constant in frequency and, therefore, highly coherent. We saw in *Plate* 2 a set of highly coherent microwave radio waves. The need for extremely high frequency stability oscillators in certain radio applications led to the development of an exceedingly narrow-band radio device utilizing atomic processes to achieve its stability. It is called the *maser,* an acronym for "*M*icrowave *A*mplification by *S*timulated *E*mission of *R*adiation." One variety, the hydrogen maser, can achieve a frequency constancy of not one part in a hundred or one part in a million, but of one part in a million million. Following the development of the microwave maser, its principle was extended to the light wave region, and light wave "oscillators" having very high coherence then became available. These are now called *optical masers* or LASERS (*L*ight *A*mplification, rather than *M*icrowave *A*mplification).

One form of laser, the gas laser, exhibits a particularly high frequency constancy and, accordingly, it is the one most often used for making holograms. Another form, the pulsed laser, provides higher intensity light but generally its light has less coherency. Earlier forms of coherent light sources (the ones available to Gabor in 1947) produce light which is less intense and less coherent than the light produced by a laser.

Spatial Coherence

Thus, with the laser, light having a high degree of frequency coherence can be generated. For holography, however, a second form of coherence is required, namely, *spatial coherence*. Many of the figures of the previous chapter showed single-frequency waves which were also very uniform. The availability of such uniform waves is just as important in holography as the availability of single-frequency waves. This requirement is particularly evident when the uniformity of the reference beam is considered.

When plane light waves from a highly coherent laser are passed through a ground glass plate, they are no longer plane waves; they are badly jumbled and mixed up. We noted that, in reconstructing a hologram, the original reference beam and the later reconstructing beam must be alike; otherwise, the hologram scene will not be reproduced faithfully. If the original reference beam used were a badly jumbled one, the rather formidable problem would exist in the reconstruction process of providing a second, jumbled-up beam exactly like the reference beam.

Coherent Waves from Small Sources

Both spatial coherence and frequency coherence are needed for holography. Fortunately, once a wave generator which can generate waves having good frequency coherence is available, the same generator can usually also form waves having good spatial coherence.

For sound waves or microwaves, spatial coherence is easily obtainable because the wave generators themselves can be made very small, only a wavelength or so across. It is seen, for example, in *Plate* 1, that the crest-to-crest distance or wavelength of the sound waves are significantly larger than the size of the telephone receiver unit from which they are issuing. Also, in *Plate* 2, the very short radio waves called microwaves are seen issuing from the small, rectangular metal tube shown at the left. In microwave terminology, this tube is called a *wave guide* and, like the telephone receiver in *Plate* 1, it, too, is smaller in size than the 1.3-inch wavelength microwaves issuing from it. Thus, both sound waves and microwaves can be radiated from sources whose dimensions are a wavelength or more in dimensions.

Plate 6 shows how the circular pattern of microwaves issuing from the wave guide at the left can be converted, by the structure immediately to its right (a metal microwave lens), into the plane wave microwave pattern displayed at the right of the lens. The photo demonstrates that coherent microwaves, like coherent sound waves, can be placed in plane wave patterns. Just as the spherical sound waves of *Plate* 1 or the spherical microwaves of *Plate* 2 correspond to the spherical light waves sketched in the hologram process of Figure 7, so the plane microwaves of *Plate* 6 correspond to the plane light waves sketched in Figure 7.

A Microwave Hologram

It is perhaps of interest to note briefly that the photographic record of *Plate* 6, first published in 1951 by the author and his Bell Telephone Laboratories colleague, Floyd K. Harvey, can be considered to be one of the early holograms. It is not a visible light hologram but a microwave hologram and was made as shown in Figure 15. A single coherent wave source (a klystron microwave generator) formed two wave sets; one was the wave set of interest (the lens waves); the

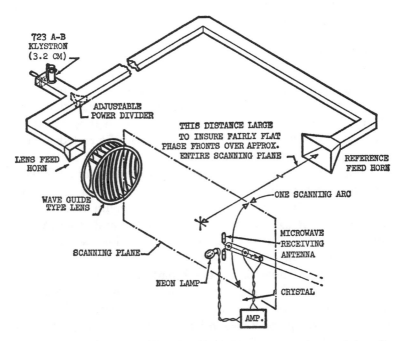

Figure 15. Two sets of coherent microwaves generated by the klystron at the upper left (one set being an off-angle reference wave) are caused to interfere at the scanning plane. The microwave interference pattern is converted into a light pattern by the scanning mechanism, and it is recorded by a camera set at time exposure.

other was a set of reference waves emerging from the horn at the right. The interference pattern generated by these two wave sets was recorded photographically in a way similar to that employed in recording the sound-wave pattern of *Plate* 1 (that photo, incidentally, can be looked upon as an *acoustic* hologram). A microwave detector was made to scan the microwave field. Affixed to it was a neon lamp whose brightness was made to correspond at every spot to the intensity of the microwave field. This varying light intensity was recorded by means of a camera set at time exposure. The picture formed is a photographic recording of the fringe pattern (the white vertical lines) generated by the interference between a set of microwaves of interest and a set of reference waves; it is, therefore, a hologram. At the time it originally was made (to show wave patterns visibly), its hologram nature was not realized.

A Real Acoustic Image

In *Plate* 7, spherical sound waves are issuing from a small horn at the left and are being redirected by an acoustic lens at the right center of the photo. The outward-spreading waves at the left are transformed by the lens into inwardly moving waves on the right; these converge at a focused area and then expand outward as though they are again emanating from a sound source, such as the horn at the far left.

The photo portrays wave actions which can be compared to several of the hologram reconstruction waves sketched in Figure 8. The converging waves immediately to the right of the lens correspond to the downward-converging waves in Figure 8, which create a real image at Pr. These converging sound waves also form a "real image"; that is, they cause a concentration of sound energy as evidenced by the white areas being thicker and brighter at the focal area. To the right of the focal area, the waves are diverging as though issuing from a new sound source. An observer far to the right would imagine that a source existed there just as an observer of the upward-diverging waves in Figure 8 would receive the impression that a light source existed at Pv. Actually, the wave concentration in *Plate* 7 is strictly classed as a *real* image, whereas the one in Figure 8 is classed as a virtual image. (We shall see later that in certain holograms the real image can be just as realistic as the virtual image usually displayed for the viewer in holography.) Above the lens, the waves from the horn are not affected by the lens, and they pass by undeviated just as the undeviated or *zero order* waves of Figure 8 pass right through the hologram.

That waves of high frequency coherence can be formed into waves having high spatial coherence is shown in *Plate* 8. The small horn on the left (it is the same horn shown in *Plate* 7) has spherical waves issuing from it, and the large acoustic lens in the center converts these into the large plane wave field at the right. *Plate* 9 shows this same lens mounted in the aperture of an acoustic horn or megaphone. There it similarly converts the diverging, spherically expanding waves within the horn into the plane waves seen at the right of the aperture in *Plate* 10. For the sound waves used, this horn-lens aperture is approximately thirty wavelengths across.

The Extremely Short Wavelengths of Light

The wavelengths of light waves are extremely small; thus, we noted earlier that violet light has a wavelength of only sixteen millionths of an inch (an optical lens 1.6 inches across would extend over 100,000 wavelengths of violet light). Because light wavelengths are so extremely small in size, it is practically impossible to construct a light generator the size of one light wavelength. Accordingly, the technique we have discussed for generating spatially coherent sound waves or radio waves—that of starting with a one-wavelength radiator and using lenses or reflectors to form spatially coherent plane wave fields—is not feasible for generating spatially coherent light waves.

Thus, two problem areas exist in achieving coherent light; one is that of achieving good frequency coherence, and the other is that of achieving good spatial coherence. Ordinary light sources, such as incandescent lamps, are "noiselike" in nature, exhibiting a very poor degree of frequency coherence. Furthermore, they possess a luminous area which is extremely large compared to one light wavelength. Even if the light from each and every tiny light generator on the incandescent surface were to be single-frequency light, spatial coherence would still be lacking because the light would be generated randomly and independently at millions of tiny points over the luminous area. Such high frequency-coherent, poor spatially coherent light would resemble laser light which has passed through a ground glass screen.

We shall see in the next chapter that, fortunately, lasers differ from previous light sources in two ways. First, the light from a laser stems from one particular atomic energy transfer process, and this causes its frequency content to be very pure (the light is single-color or *monochromatic*). Second, it employs reflecting surfaces in the light generation process which cause the light to be emitted in the form of extremely plane waves whose wave fronts are many, many wavelengths across. Lasers thus have made available light having both good frequency coherence and good spatial coherence.

Chapter 3

LASERS

Several classes of lasers are now available. They include the pulsed or ruby laser (often called the solid state laser), the gas laser, and the semiconductor or injection laser. The laser process was first achieved and demonstrated with the first type; it also produces the most powerful light pulses.

The First Laser

We noted earlier that the earliest stimulated emission device to be developed was the microwave maser, first demonstrated in 1945. For its development, the Nobel Prize in Physics was awarded jointly in 1964 to the U.S. scientist, Charles H. Townes, then at the Massachusetts Institute of Technology in Cambridge, Massachusetts, and to the Soviet scientists, A. M. Prokorov and N. Basov, both at the Lebedev Institute in Moscow. Following the success of the maser, many workers endeavored to extend its use from microwaves to light wavelengths. In 1960, the U.S. scientist, T. H. Maiman, then at the Research Laboratories of the Hughes Aircraft Company in California, demonstrated the first laser, using a ruby rod as the active element. His original laser is shown in *Plate* 11. Since the basic process for generating light is common to all three types of lasers, let us examine the workings of Maiman's ruby laser in some detail.

The active material, which can be ruby or various kinds of especially "doped" glass, is shaped into a cylindrical rod, as shown in Figure 16. Around this rod is wrapped a helical flash tube which, when connected to a powerful source of stored electrical energy, emits a very short and intense burst of broad band (incoherent) light. Some of this light energy is absorbed by the atoms of the rod, and in the process, the atoms are excited, *i.e.,* they are placed in an *energy state* which is at a higher *energy level* than the state in which most of them "reside." Energy is thereby stored in these atoms, and when they return to their normal unexcited state (we shall call this the *ground state*), they release their stored energy in the form of light waves.

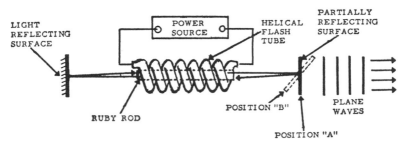

Figure 16. The essential parts of Maiman's ruby laser. (For simplicity in the text, the reflectors are here shown separated; they usually are provided by silvering the flattened ends of the ruby rod.)

Up to this point, there is nothing special about the light-emitting process just described; practically all light sources generate their light in this way. The atoms in an incandescent lamp filament are excited to higher energy states by the heat energy provided by the electric current passing through the filament, and as these atoms return to lower energy states (only to be again excited), they release this energy difference in the form of light.

The Danish scientist, Niels Bohr, suggested many years ago that the radiation of spectral lines by atoms could be explained by assuming that electrons revolve about the atom's nucleus in certain fixed orbits (like the planets circle the sun) and that each of these orbits represents a definite energy level. When an electron is in an outer orbit, the atom is in a state of higher energy, an excited state, and when the electron transits to an inner orbit, energy is radiated in the form of spectral lines that are characteristic of a particular atom.

Bohr thereby obtained values for the spectral series of visible hydrogen lines (colors) with an accuracy which was quite astonishing.

The significant difference between the laser and other light sources is that the laser light source material provides a particular form of energy state in which the excited atoms can and do pause before returning to their ground state. They tend to remain in this state (called a *metastable* state) until stimulated into returning to the ground state. In this last step, they emit light having exactly the same wavelengths as the light which triggered them into leaving that state. The atoms are thus *stimulated* into *emitting*—hence, the words "stimulated emission" in the laser acronym. Stated another way, energy is first stored in the atom and later released by it when it transfers from the metastable state to the lower one in the form of single wavelength light energy.

A Water Wave Analogy

This atomic process can be compared somewhat crudely, as shown in Figure 17, to that of lifting a heavy object, such as a spherical rock, and placing it on a shelf located above the surface of a body of water. This shelf is one which an oncoming water wave can tilt and thereby cause the rock to fall back into the water. When this happens, wave action is

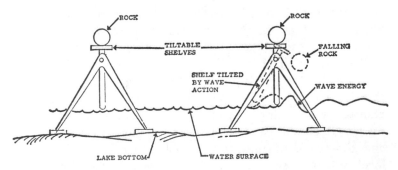

Figure 17. Rocks placed on shelves which can be tilted by water waves provide a crude analogue to the laser's metastable state. The water waves cause the rock to leave its elevated energy state and thereby generate more water waves.

produced. In the process of lifting the rock to the shelf, energy (potential energy) is stored in the rock. When the rock later rolls off the shelf, this energy is transformed into kinetic energy, and much of it is transferred to the water waves which are generated by the rock striking the water. This energy travels with the new water waves as they move outward. Should there be, as shown at the left, a second shelf with a similar rock on it, in the path of these waves, the new rock-generated waves can cause this second rock to fall in the water, thereby generating still more waves.

In this analogy, the rock resting on the ground or lake bottom corresponds to the laser atom in its low energy state, the ground state. The placing of the rock in its shelf position corresponds (but only approximately) to placing the atom in its excited, metastable state. The water waves which tilt the shelf and cause the rock to fall correspond to the light waves which *stimulate* the atom into falling into its lower energy state, thereby emitting light waves. The rock similarly falls back into the water (into *its* ground state) and generates water waves. These new water waves then cause the second rock to fall off its shelf (thereby generating more water waves) and, hence, correspond to the light waves generated by the falling atom which similarly stimulates other atoms into falling and thereby emitting additional light.

The Two-Step Process

One difference between the laser and the water wave cases is that, in the laser, the emitted light waves are of exactly the same frequency as those which stimulated the atom to emit. These new waves are thus exactly suited to react with other metastable atoms and cause them to emit more of this same radiation. We also noted that the lifting of the rock to its shelf is only approximately equivalent to placing the atom in its metastable state. This is because the usual laser process is not a one-step process such as that of placing the rock on its shelf; it is, instead, a *two*-step one, as the energy diagram of Figure 18 indicates. Such energy diagrams, representing atomic processes, are plotted vertically with the lower energy states placed low on the vertical scale, parallel with the situation where objects at lower heights have lower (potential) energy. The lowest level, the ground state, represents that energy level to which atoms in the metastable state transfer. The excitation energy of the flash tube imparts to atoms residing in the lower level sufficient energy to raise them to the energy state represented by the highest of the three levels shown. From this level, they fall to the middle level state—the metastable state. The nature of this metastable level is

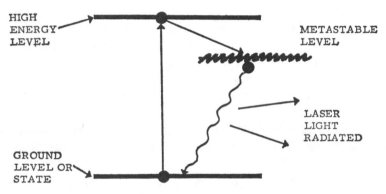

Figure 18. Three energy levels are available to atoms or ions in a laser material. Transition from the middle, metastable state to the ground state results in the emission of laser light.

such that the atoms tend to remain in it. When, however, the exactly correct wavelength light impinges on an atom in this state, it will depart from this state and fall to the lower energy state, emitting, in the process, a burst of energy in the form of light. The light burst which accompanies its return to the lower level can cause further emission of this same wavelength radiation from other atoms residing in the metastable state.

We see, therefore, that laser action depends upon the existence of this special state in the laser material. The process of placing, by means of the high energy flash lamp, a large percentage of atoms in the laser material in the metastable state is referred to as a *population inversion,* because, originally, most of the atoms reside in lower energy states. After the flash, an increased number of atoms reside in the metastable state.

Just how extraordinary can a metastable state be? As a general rule, the time that an atom spends in a *normal* excited state (this time is referred to as the *lifetime of the excited state*) is on the order of one hundred millionth part of a second (10^{-8} seconds). In contrast, there is one metastable state which has an average lifetime of almost a full second (100 million times longer than that of the average excited state). This state is involved in the generation of the green line of atomic oxygen (at a wavelength of approximately 22 millions of an inch) observed in the luminous night sky phenomenon called the *aurora*. (The name *aurora borealis*, meaning "northern dawn," was proposed in 1620 by the French philosopher, Pierre Gassendi.)

Energy Conservation with Reflectors

Although our earlier analogue of the rocks and tilting shelves is rather crude, it can, nevertheless, be extended one step farther to help explain the use of reflectors in lasers. When water waves travel outward in all directions from a point, their energy remains distributed over the constantly increasing circular wave perimeter, and, consequently, the wave height (wave energy) continually decreases. If this circular spreading is permitted, the ability of these new water waves to tip the second shelf diminishes with increased distance between the two shelves. A similar situation exists for the case of the freely radiating light waves which issue from an atom which has been stimulated to emit radiation.

When, on the other hand, the newly generated water waves are *confined*, such as, for example, in a bath tub, they are *reflected* by the walls, and the wave energy, instead of spreading out, is conserved. The use of reflecting surfaces in lasers similarly conserves the desired light wave energy. Such surfaces cause a reflection of light wave energy to occur, and this makes more available to other metastable atoms the desired special wavelength energy. This energy can then continue to stimulate more atoms into emitting more of this special radiation.

This use in a laser of reflecting surfaces to conserve the desired light energy is shown in Figure 16. Some of the light generated by metastable atoms returning to their lower energy state will travel down the ruby rod and be reflected by the reflecting surface, here shown for simplicity as separated from the rod itself. (In most ruby lasers, the ends of the rod are coated, and they act as the reflecting surfaces.) Much of this light will be reflected back into the rod, and thus it can stimulate other atoms into emitting more of the desired light energy. The stimulated-emission process builds up, and the result is that huge numbers of atoms soon radiate one powerful pulse of laser light.

Unless the conservation process of reflection is provided, lasers often are unable to oscillate, that is, to emit light. For many of the most important applications of the microwave maser, on the other hand, the oscillating or wave-radiating condition is not desired. Masers are used primarily for amplifying weak radio signals, and, therefore, only enough stimulated emissions are permitted to provide amplification, not enough to cause uncontrolled oscillation. For lasers, the greatest interest at the present time is in their light wave generation capabilities. Some day, perhaps, the equally possible *amplifying* capability of the laser also will become important. Many envision the day when light beams propagating in hollow tubes will carry much of the world's communications, as radio waves propagating in cylindrical, coaxial *cables* now carry telephone and television signals. In such an application, the light beams would be varied in intensity (*modulated*) at an extremely high rate, thereby being *coded,* as a telegraph signal is now coded (at, however, a very *low* rate). This coding could permit the transmission, via light beam, of millions of simultaneous telephone conversations and many thousands of simultaneous television signals. After such a coded light beam would travel some distance in the tube, it would decay in intensity, and the amplification capability of the laser then would permit the light beam and its signals to be amplified back up to their original strength.

PLATE 1. Single wavelength sound waves are portrayed here issuing
from a telephone receiver. They spread out in a spherical pattern,
the cross section of which resembles the water wave pattern of
Figure 1.

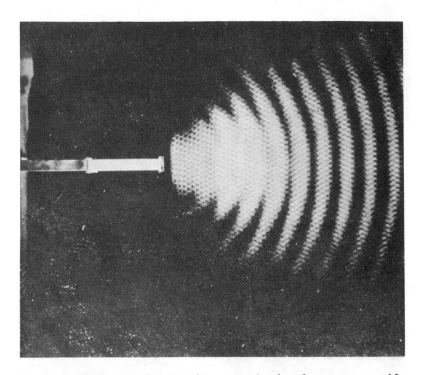

PLATE 2. Single wavelength microwaves issuing from a waveguide. Microwaves, like light waves, are electromagnetic waves.

PLATE 3. An interference pattern formed by two sets of coherent sound waves.

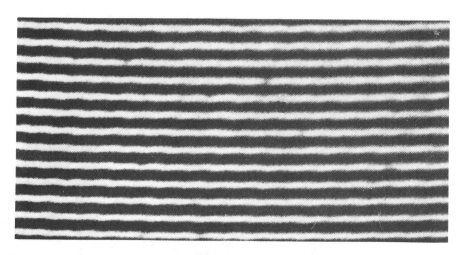

PLATE 4. At the plane of the photographic plate of Figure 5, the wave interference pattern is a series of bright and dark lines. This figure is a photographic recording of such a pattern.

PLATE 5. At the plane of the photographic plate of Figure 7, the wave inteference pattern is a series of bright and dark circles. This figure portrays a photographic recording of such a pattern.

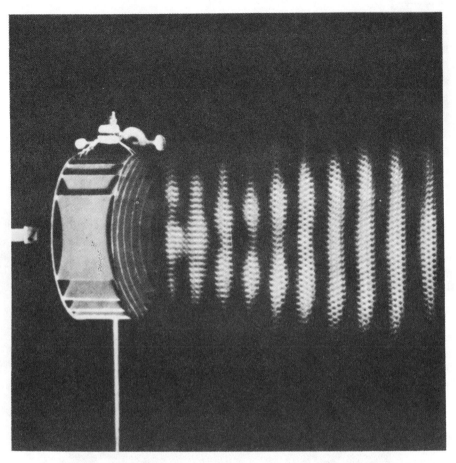

PLATE 6. The spherical waves issuing at the left from the same wave-guide as in *Plate 2* are converted into plane waves by a microwave lens. This photo, made as indicated in Figure 15, is a recording of the interference pattern formed by a wave set of interest and a reference wave; it can accordingly be considered to be a microwave hologram, the white vertical striations being microwave *fringes*.

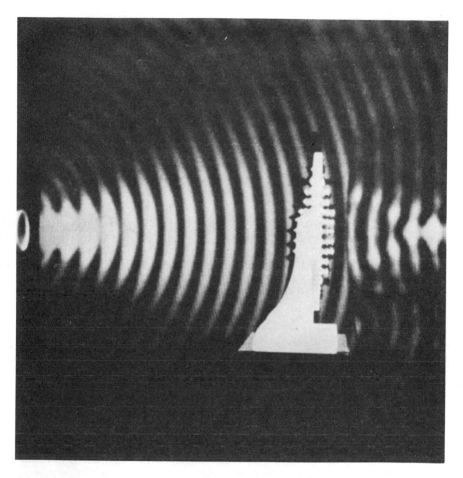

PLATE 7. Circularly diverging sound waves issuing from the horn at the left are converted, by the acoustic lens, into circularly converging waves at the right. After they pass through the focal point, the waves again diverge (at the far right).

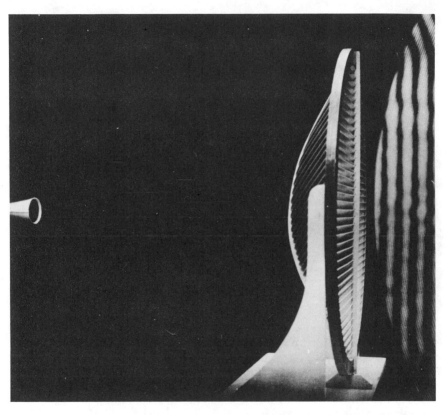

PLATE 8. An acoustic lens converts into plane waves the circular wave fronts of sound waves issuing from a horn placed at the focal point of the lens.

PLATE 9. An acoustic lens can be mounted in the aperture of a coni-
cal horn.

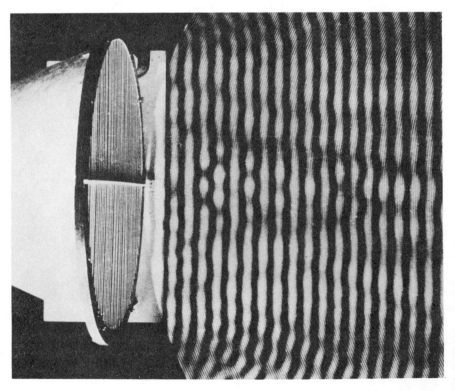

PLATE 10. This photograph shows the wave-beaming effect of the lens in *Plate 9*.

PLATE 11. A photograph of the first laser. It was designed by T. H. Maiman at the Hughes Aircraft Company.

PLATE 12. A steel razor blade disintegrates when the coherent light waves from a ruby laser are focused sharply on it. (Courtesy Bendix Aerospace Systems Division)

PLATE 13. Discoverer of the effect utilized in some lasers, Nobel Laureate Sir C. V. Raman (standing, center) poses with a group of Bell Laboratories scientists. Among them are Nobel Laureates Walter Brattain (seated beneath Raman) and John Bardeen (second from left, standing).

PLATE 14. Partially illuminated by the sun off to the right, the earth appears as a crescent to U.S. moon probe, NASA's Surveyor VII. Directly above the arrow are two tiny spots of light formed by laser beams aimed toward the moon and originating at Kitt Peak, Arizona (right spot) and Table Mountain, California (left spot). (Courtesy National Aeronautics and Space Administration)

PLATE 15. This amplitude pattern of coherent sound waves issuing from the thirty wavelength aperture horn-lens of *Plate 9* indicates that the beam remains collimated (parallel sides).

PLATE 16. Two separated sound sources act like two optical slits. A diffraction pattern of constructive and destructive interference is created. The horizontal, zero-order component is evident; also evident are the upward and downward diffracted first order components, and the fainter second order components.

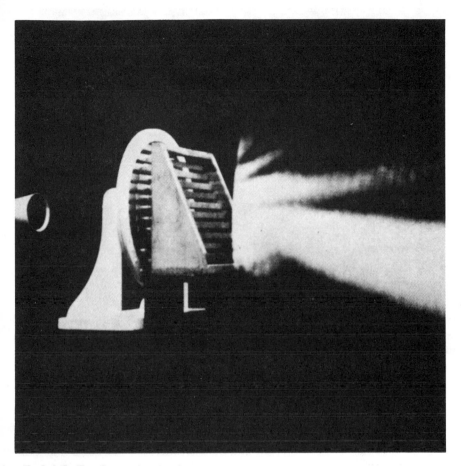

PLATE 17. The focused beam of sound waves from an acoustic lens is deflected downward by an acoustic prism.

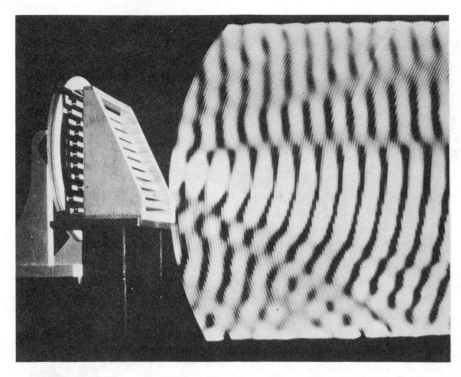

PLATE 18. This photo shows the wave pattern of the deflected sound energy portrayed in *Plate 17*.

Reflectors and Spatial Coherence

In addition to conserving energy, the reflectors also enhance the spatial coherence of the laser light. We recall in the discussion of coherence that good wave *planarity* was equivalent to good spatial coherence. In the case of the sound waves in *Plate* 8, this planarity was achieved easily with a lens because the wave source was very small, comparable to a wavelength in size. We noted, however, that such small sources are not possible at light wavelengths and that good planarity for light waves must be achieved in some other way. In the laser, the reflectors provide the means. When the single-frequency waves generated by the laser are made to bounce back and forth between two accurately made reflecting surfaces, they acquire, even after a few traverses, a very high degree of planarity. As shown in Figure 16, one of the reflectors (at the right) is made slightly transparent, so that not all of the energy striking it is reflected. Thus, the light which does pass through is the usable, radiated, laser light.

Energy Concentration

In Figure 16, the light waves emitted at the right from the laser are shown as a progression of plane waves having high *spatial* coherence; in addition, the nature of the light generation ensures that these waves have high *frequency* coherence. Having, thus, *full* coherence, these waves can be manipulated in exactly the same way that coherent sound waves or coherent microwaves can be manipulated. For example, if a lens is placed immediately to the right of the plane wave area at the right of Figure 16, it will focus the highly coherent plane laser waves into an extremely small volume. The situation would be similar to the lens of *Plate* 7 focusing the sound waves in that figure into a small region. A closer comparison would be to consider this focusing of the laser light by a lens as equivalent to the lens action of Figure 12 *reversed*. In that figure, the waves are originating at the small horn aperture at the left. However, one could assume that the plane sound waves at the right are arriving *from* the right. In progressing toward the left through the lens, these plane waves would be converted into spherical waves and concentrated into a small volume comparable to the dimensions of the horn aperture (a wavelength or two across).

Thus, when an optical lens is placed at the right of Figure 16, the laser light similarly can be concentrated into a volume perhaps only a wavelength or two across, and because light waves are only a few millionths of an inch in length, the power flow per unit area in this focal region can become fantastically high.

An idea of the enormity of this energy concentration can be had by calculating what the power flow would be if all the energy radiated by a 75-watt light bulb were made to pass through a square area one wavelength of violet light (16 millionths of an inch) on a side. This would be equivalent to a power flow, per square *inch*, of approximately 300,000 million watts, *i.e.*, equivalent to funneling this enormous amount of power through a one-square-inch area. This amount of *electric* power is larger than that which at the present time could be generated by *all* the electric power stations in the entire United States operating simultaneously. Even the very first lasers could manifest their tremendous energy-concentration capability in numerous spectacular ways. For example, laser light, when sharply concentrated, can punch holes in steel razor blades, as is shown in the 1963 photo of *Plate* 12; here, the flying sparks add to the spectacle.

Q-Spoiling

We mentioned earlier that lasers generally do not oscillate (emit light) when the reflectors are absent. Also, the laser will not oscillate if the two reflectors are not positioned properly. This situation has made possible the obtaining of extremely powerful laser pulses. If the right-hand reflector in Figure 16 is tilted (as shown by the dotted lines) so that it no longer reflects energy back into the laser rod, the normally desired energy conservation process is negated. However, the flash lamp pulse, during its very brief lifetime, continues to place more and more atoms into their metastable state. Because very few are stimulated into falling back to the ground state, a much larger population inversion is achieved; many atoms are elevated to their metastable state, and few can leave it. If, now, at the proper instant, the reflector is rotated into its proper energy conserving condition, a very sudden, almost avalanche-like transition of atoms takes place from the metastable state to the ground state, and a very short and extremely powerful pulse of single-frequency light is generated.

This process of negating the energy reflection or conservation procedure in lasers is called "Q-spoiling." The term arose from an expression used in radio engineering. In early radio sets, the tuning circuit comprised an inductance and a condenser, and the quality or sharpness of tuning of this circuit was specified by the term, "Q," the first letter of the word quality. A sharply tuned (high Q) circuit passes a very narrow band of frequency (it could, for example, in Figure 14, cause the one percent band shown to become appreciably narrower). This higher quality is usually achieved by reducing circuit losses. If the losses are too high, that is, if the Q of the circuit is too low, a radio oscillator employing such a tuned circuit cannot oscillate. Directly analogous with this inability of a radio oscillator to oscillate because of the poor Q in its tuning circuit, the tilting of a laser mirror to inhibit the laser from oscillating is now referred to as *Q-spoiling*. When the mirror becomes properly positioned (so that the Q of the laser circuit is no longer "spoiled"), oscillation occurs.

Gas Lasers

The second form of laser to be developed was the gas laser. It consists most simply of a glass tube filled with a special gas mixture. High voltage is applied across two electrodes near the ends of the tube, as shown in Figure 19, causing an electrical discharge to take place. The gas glows and the tube looks much like the glass tube of an ordinary neon advertising sign.

The gas laser differs from a neon sign in that its gas mixture provides the necessary metastable state in which the excited atoms can temporarily reside. As in the ruby laser, the energy difference between the metastable state and the lower state to which they fall corresponds to the energy of the single color light which is radiated. The first gas laser used a mixture of helium and neon and, like the first ruby laser, generated red light; its light had, however, an orange red color, rather than the crimson red of the ruby. Its wavelength was 63.3 millionths of a centimeter (6328 Angstrom units) whereas the light from the ruby laser has a wavelength of 69.3 millionths of a centimeter (6930 Angstrom units). Other

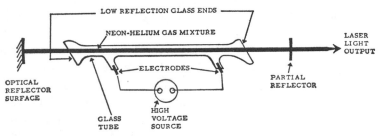

Figure 19. Gas lasers often use a direct-current glow discharge for exciting the atoms or ions to the required higher energy levels.

gases are now used, argon providing blue-green (4880 Angstrom units) and green (5145 Angstrom units) laser light, nitrogen providing 3371 Angstrom units, and carbon dioxide providing infrared laser light (10,600 Angstrom units).

One significant difference between gas lasers and ruby lasers is that the gas laser operates continuously. The glow discharge caused by the applied voltage continually places a large population of atoms in a metastable state, and although many atoms constantly are falling back to a lower level, many continually are elevated again by the glow discharge phenomenon.

Reflectors and Resonators

The ends of the glass tube of a gas laser are usually fitted with special low-reflecting glass surfaces, and the needed reflecting surfaces are placed externally to these, as shown in

Figure 19. For most gas lasers, the active portion (the gas tube) is much longer than the ruby rod of a pulsed laser. This feature, along with the feature of continuous operation, significantly improves the coherence of their light. The two reflectors in a laser act as the end walls of a resonator, and it is this greater distance between reflectors which enhances the frequency coherence. To clarify this point, let us review certain properties of resonators.

A piano string is a resonant device. It can vibrate either at its fundamental frequency or at àny multiple of that fundamental. In like manner, a closed box (or resonator) can vibrate or resonate at its lowest frequency or at any one of its higher frequency or higher-order *modes*. This is illustrated in Figure 20. Here the box is small in cross section and resembles the resonator of a wooden organ pipe. At the top of the

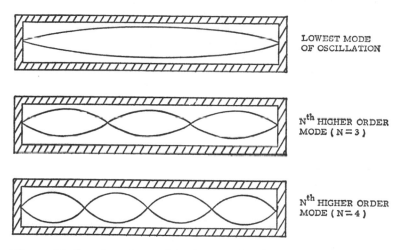

LOWEST MODE
OF OSCILLATION

N^{th} HIGHER ORDER
MODE ($N = 3$)

N^{th} HIGHER ORDER
MODE ($N = 4$)

Figure 20. In a long box with closed reflecting ends, sound waves can resonate at various frequencies, as in an organ pipe.

figure is sketched the wave pattern of the lowest or fundamental mode of oscillation. Reflection occurs at each end and maximum air motion exists at the center. In the center is sketched the third order mode; this oscillation has a frequency three times that of the oscillation at the top; it is said to be its third harmonic. At the bottom is sketched the fourth-order ($n=4$) mode; it has one-fourth the wavelength and four times the frequency of the fundamental. For this wave, the length of the resonator is seen to be four half-wavelengths (or two full wavelengths).

Imagine the four side walls of the wooden resonator to be removed and its two reflecting ends replaced with accurately made mirrors placed eight inches apart. The separation between them then would be, for violet light (whose wavelength is 16 millionths of an inch), one million half-wavelengths. Accordingly, if violet light is resonating between these two mirrors, a change to the next resonant mode, up or down, would result not in the three-to-four change of Figure 20, but in a far smaller percentage change, a change of only one part in a million.

In Figure 21, a frequency plot is drawn for the three cases of Figure 20. The two center drawings represent the higher modes corresponding to $n=3$ and $n=4$. In the lowest sketch,

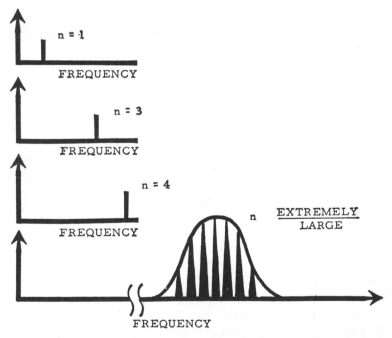

Figure 21. The top three sketches are frequency analyses of the three wave resonances of Figure 20. The fourth analysis portrays the extremely high-frequency light-wave resonances which are possible with widely spaced laser reflectors.

an extremely large increase in frequency is indicated by the break shown at the left in the horizontal frequency scale. That sketch is intended to represent a condition similar to that of the two-mirror violet light resonator just discussed; let us assume that each frequency indicated differs from its adjacent one by one part in a million. If the violet light reflecting back and forth between these two reflectors could be forced to resonate in only one mode, its single-frequency qualities (its frequency coherence) could be described as having a constancy or stability of one part in a million.

The Natural Line Width

A smooth curve also is indicated in Figure 21. We stated earlier that the natural atomic process by which laser light is generated is an important factor in providing the single frequency characteristic. However, even this constant-energy process does not produce exactly single-frequency light. The observed line width is *broadened* by two important effects, one is called Doppler-broadening and the other is called relaxation-broadening; both are traceable to the random motion of the emitting atoms. Thus, although the energy process itself corresponds to an extremely narrow spectral line, the radiated line we observe has a significant frequency spread; this width is usually referred to as the *natural line width*. The smooth curve or envelope of Figure 21 is intended to indicate this natural line width.

We recall that the Doppler effect is noticeable as a drop in frequency or pitch of the whistle of a fast locomotive as it passes a listener. As the source approaches, the sound possesses an "up-Doppler," and as it recedes, a "down-Doppler." To the locomotive engineer, however, no frequency change is observed. For a moving atom, the frequency of its radiated light may similarly be quite constant and precise, *relative to the atom*, but to an observer located along the line of its motion, the light frequency would be Doppler-shifted.

An interesting use of artificially imparted Doppler shift to laser light was suggested recently by the U.S. scientist, Richard Milburn, at Tufts University. In his procedure, laser light is reflected from a beam of rapidly moving (high energy) electrons issuing from an electron accelerator. Just as a slowly moving billiard ball has its direction reversed and its velocity (energy) greatly increased when struck by a second, oncoming, rapidly moving billiard ball, so the reflected laser light waves acquire a very large Doppler increase in their frequency. Higher *frequency* light means higher *energy* light, and in this process, the light is shifted so drastically in frequency that the laser beam is transformed into a beam of high energy x-rays (actually *gamma* rays). Furthermore, these gamma rays have properties virtually impossible to obtain with ordinary gamma ray sources. First, the laser light can be *polarized* (by passing it through the glass from a pair of polaroid sun glasses), which causes the resulting *gamma* rays to be polarized (a very unusual situation). Second, the monochromaticity of the laser light (high frequency coherence) results in the gamma radiation similarly being quite monochromatic (also unusual).

Recently some atomic transitions have been observed for which the natural line width is quite narrow. One procedure for achieving this was discovered and utilized by the German scientist, Rudolf Mossbauer, in 1957 (he received the Nobel Prize for his work in 1961). He employed a radiating substance in which the atoms were so firmly bound in a crystal lattice that they could not move, even when they radiated. The elimination of motion in these atoms caused their radiation to be extremely frequency coherent. Unfortunately, this procedure cannot be used in lasers; for them, a sizeable natural line width is always observed.

Fortunately, however, the individual resonator modes of a laser can be appreciably narrower than the natural line width of the laser, and hence, this narrower width provides a significant improvement in frequency coherence. There often is a problem in keeping the oscillation in only *one* of these numerous modes. Many lasers oscillate in several modes simultaneously, and for others, mode "jumping" is constantly occurring. These phenomena obviously impair the coherence of the laser light. To minimize the mode-jumping effects, extremely high mechanical stability is usually employed. Thermal contraction of the reflector supports can result in a short-ening of the distance between reflectors, thereby causing the oscillator frequency to change from one resonator mode to another. When, however, extremely good mechanical and thermal stability is maintained, it is possible for the laser to operate in one single mode of oscillation and thereby possess an extreme degree of frequency coherence.

Coherence Length

Because gas lasers provide excellent coherence, they are the lasers most commonly used in making holograms. But even here some gas lasers are better than others. For holography, the most pertinent way of defining the extent of a laser's frequency coherence is by specifying its *coherence length*. Let us examine what this term implies.

We noted earlier that interference effects still occur for single wavelength waves even when one of the wave sets is slid an integral number of wavelengths ahead or behind the other set. This, of course, assumes that the separation between wave crests remains identical and that each wave crest and trough is exactly like all succeeding crests and troughs. Actually, we find that after one set has been shifted relative to the other by many, many wavelengths, a point is eventually reached where the addition and subtraction effect is no longer perfect. It is easy to see that this is a real problem at light wavelengths. When two wave sets of plane, coherent, violet light waves are shifted only sixteen inches, this shift is one encompassing one million wavelengths. A wavering in frequency of only one part in a million would cause these two crests to be out of step.

The distance over which laser radiation *does* remain in accurate step is called the coherence length; some gas lasers exhibit coherence lengths as large as several meters. For use in holography, lasers having such long coherence lengths permit the recording of scenes having much greater depths than could be recorded with shorter coherence length lasers.

Semiconductor Lasers

The third form of laser to be developed was the semiconductor or injection laser. It is similar to the ruby and gas laser except that it employs as its active substance a tiny piece of semiconductor material. A semiconductor is a material which is neither a good conductor of electricity nor a perfect insulator. When used as a laser, direct current is made to flow through the material by connecting a power source to two electrodes affixed to opposite surfaces. Atoms within the material are thereby excited to higher energy states, and in falling back to a lower state, give off light or other forms of electromagnetic radiation. Solid state lasers are particularly useful for generating radiation in the infrared region and in the millimeter radio wave region. They have not been used appreciably up to now in holography.

Laser action has been produced in certain substances (including liquids) by utilizing a light-scattering process called, after its discoverer (the Indian scientist, Sir C. V. Raman), the Raman Effect. This effect is observed when light waves falling on a material interact with the internal vibrations characteristic of the material, thereby causing sum and difference frequency light waves to be generated.

In some arrangements, a resonator is made of the Raman material (such as a block of quartz having plane parallel-end faces) and a powerful ruby laser beam serves as the means for exciting the *lattice* vibrations in the material. Thus, just as the flash tube is the "pump" or energy source, for the ruby laser, a laser beam here acts as the "pump." The intense lattice vibrations generated in the quartz *modulate* the ruby light beam, thereby generating the upper and lower *sidebands* (sum and difference frequencies). Parallel with the term, "stimulated emission," this process is referred to as "stimulated Raman scattering" (SRS). Raman received the Nobel Prize in Physics for this effect in 1930. *Plate* 13, taken at the author's home, includes Sir C. V. Raman and two other Nobel Laureates.

Chapter 4

WAVE DIFFRACTION

In describing the hologram process in the first chapter, the terms "deviated upwards" and "diffracted upwards" were rather loosely interchanged. Because the phenomenon of diffraction is very important in holography, we will now look at it more rigorously and in greater detail.

Webster's dictionary defines diffraction as "A modification which light undergoes, as in passing by the edges of opaque bodies or through narrow slits, in which the rays appear to be deflected, producing fringes of parallel light and dark or colored bands; also the analogous phenomenon in the case of sound, electricity, etc." From this definition, we see that there are several aspects of diffraction. We shall discuss first the diffraction effects caused by one or more slits.

Diffraction by a Slit

Let us consider the case of a distant light or sound source illuminating a slit in an opaque screen, as shown in Figure 22. What is the radiation pattern which is formed by the slit in the dark area (the shadow area) behind the screen? In Figure 23, this illumination pattern, as cast on a second screen, is sketched. It has a bright central area flanked by a series of maxima and minima. The pattern is plotted in more detail in Figure 24. One property of this pattern, the width of the beam created at great distances by the waves passing through the slit, is rather interesting. This beam width is inversely proportional to the aperture or slit width a and directly proportional to the wavelength λ; when a and λ are expressed in the same dimensional units, this width, in angular degrees, is 51 λ /a.

Let us see what sort of beam width this figure of 51 λ /a gives in various situations. We calculate first the theoretical

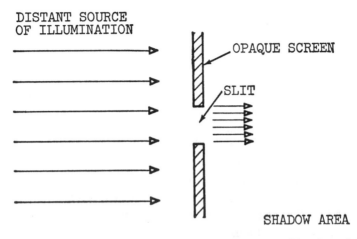

Figure 22. When a slit in an opaque structure is illuminated by a distant source of light, the energy distribution is uniform over the width of the slit.

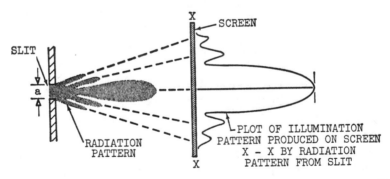

Figure 23. In the shadow region behind the opaque screen of Figure 22, light emerges from the slit in a beam-shaped pattern.

beam width of the 200-inch optical telescope at Mount Palomar Observatory in California. In Chapter 1, we noted that violet light has a wavelength of about sixteen millionths of an inch. For this violet light and the 200-inch mirror, $51 \lambda/a$ becomes $51 \times \dfrac{16}{1,000,000} \times \dfrac{1}{200}$ degrees, or approximately four millionths of a degree. This calculation says that if the 200-inch telescope is energized uniformly over its aperture with coherent violet light, the beam formed by it is but 370 feet broad at a point one million miles away.

Let us calculate, as a second example, the spread which a ruby laser beam might experience in going from the earth to the moon. Assume that the cross-sectional dimension of the ruby rod is one-half inch, that is, that its aperture is one-half inch across. Further assume that its radiation is fully coherent

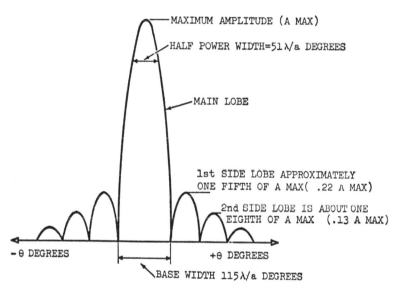

Figure 24. The diffraction pattern formed by a slit.

and that it has a wavelength, in the red light region, of thirty millionths of an inch. The formula $51 \lambda/a$ then gives a beam width of $51 \times 2 \times 30 \times 10^{-6}$ degrees, or about three one-thousandths of a degree. Accordingly, by the time this one-half inch beam has reached the moon, it would have spread in width from its original one-half inch to about twelve and one-half miles.

In a recent NASA experiment, two relatively low power

lasers, located in the Western U.S., were aimed at the moon shortly after a Surveyor moon craft had landed on the moon. The Surveyor craft took a photo of the earth when that part of the earth in which the lasers were located was in nighttime darkness, and it relayed this photo back to earth. It is shown in *Plate* 14. Two dots of light (arrow) are evident at the positions where the lasers were located; yet, the nearby brilliantly lighted cities of Los Angeles and San Francisco are not visible. This experiment demonstrates one significant advantage of spatially coherent waves. For them, an aperture many wavelengths across provides a directivity, a power gain, which can only be duplicated, for incoherent waves, by a huge increase in their radiated power. It is estimated that, at the moon, the two or three watts of light power emitted by each of these two lasers were equivalent, because of the directivity made possible by the coherence of their laser light, to ten *billion* watts of power of incoherent light. It is perhaps a little surprising that coherent light beams spread out so very little in traveling great distances. We do not observe such strong directional effects with water waves and sound waves. The sound waves from the telephone receiver portrayed in *Plate* 1 are spreading out uniformly in all directions; for them, there is no beam effect at all. An acoustic horn or megaphone does exhibit some directionality for sound waves but not a very appreciable amount. Inherent in the factor 51 λ /a is the fact that only for apertures which are many wavelengths across is the distant beam width narrow.

The Near Field

In holography, we are more interested in diffraction effects produced nearby rather than at great distances. Near the slit of Figure 22, the light pattern remains the same width as the width of the slit; that is, the beam remains completely collimated in this region. The distance from the slit for which the beam dimension matches the aperture dimension is, again, a function of aperture size and wavelength. This distance is equal to the slit width or aperture dimension squared, divided by twice the wavelength. Thus the beam from an aperture thirty wavelengths across will itself remain thirty wavelengths across out to a distance 450 wavelengths from the aperture. *Plate* 15 shows sound waves being radiated from a horn-lens combination whose aperture is thirty wavelengths across. This photo portrays only the amplitude of these sound waves; a similar photo showing the wave pattern is *Plate* 9. Although *Plate* 15 only shows the beam out to a distance of perhaps thirty or forty wavelengths, it is seen that the edges of the pattern remain parallel, *i.e.,* the width of the beam remains the same as the aperture width.

Diffraction by Two Slits

We now consider the diffraction effects which occur when light passes through two slits in an opaque screen. Numerous very bright and very dark areas result, and these are called, in optical terminology, *fringes.*

The two-slit effect of optics can be simulated with two non-directional coherent sound sources, as shown in *Plate* 16. This figure portrays the fringe pattern formed by two sources separated by three wavelengths; in this photo, the wavelength of the sound waves is approximately one and one-half inches. The combination of two identical wave fields results in wave addition and wave cancellation. As would be expected, wave addition occurs at points equidistant from the two sources, that is, along the center line of the two radiating points; this is the central, bright, horizontal area. Wave cancellation (destructive interference) occurs at those points where the distance from one source differs from that to the other source by one-half wavelength. At such points, one of the two wave sets has crests (positive pressures) where the other has troughs (negative pressures). The two areas where this half-wavelength destructive interference effect is evident in *Plate* 16 are the black areas immediately above and below the central bright area.

Bright areas are again seen above and below these two black areas. These are areas where the distances from the two sources differ by one *full* wavelength. One of the wave sets is a full wavelength ahead of the other and wave crests and wave troughs again coincide, so that positive pressures add to form higher crests and negative pressures add to form deeper troughs. Similarly, in those areas, where one of the two wave sets is two wavelengths ahead of the other, wave addition again results; these are the shorter bright areas at the very top and very bottom of the figure.

If the distances differ by an odd number of half-wavelengths, destructive interference occurs. We have already noted the dark areas where the distances differ by one half-wavelength; a second pair of dark areas are seen at the upper and lower parts of *Plate* 16; these are directions for which the two wave sets differ by three half-wavelengths.

Diffraction by Multiple Slits

If the waves portrayed in *Plate* 16 were single-wavelength *light* waves rather than sound waves, and if the radiators were long slits perpendicular to the paper, a white screen placed to the right of these slit radiators would display a comparable series of bright and dark bands, or fringes. We indicated earlier, in connection with the photographic grating of Figure 6, that when the two wave sets involved are a wavelength behind or ahead of each other, the waves add, and wave energy is diffracted in those off-angle directions for which such a condition holds. We see in *Plate* 16 exactly this effect. The only difference is that this photo portrays the diffraction pattern of two open spaces instead of for the entire array of open striations of the grating. The action of the grating is identical, however, since all the slits contribute to the wave energy diffracted in these off-angle directions in the same way as the two radiators of *Plate* 16.

All three of the wave sets discussed earlier in connection with a grating are seen in *Plate* 16. The undeviated, zero-order waves manifest themselves as the horizontally directed white area, the first-order diffracted waves (having a one-wavelength slippage) are the first upward tilted white area and the first downward tilted white area. Also evident in this photo are two still more widely deflected wave sets (the smaller, more widely diverging, white areas); these correspond to a slippage of *two* wavelengths and are called *second-order* diffracted sets.

Multiple Slit Gratings and Photographic Gratings

Although the two sound-source pattern of *Plate* 16 is representative of the pattern generated by light waves passing through two narrow slits in an opaque screen, and although this analogy can be extended to the case of an opaque screen having many equally spaced slits, we must not conclude that the earlier-discussed, photographically recorded interference pattern or grating performs in exactly this same way. A photographically made grating differs in several ways from the simple multiple slit grating we have been discussing. Let us examine Figure 25. We see that in the slit grating, the slits are completely transparent and the portions between the slits are completely opaque. In the photographic recording, the change from transparent to opaque area is gradual, rather than sudden. This causes the light amplitude in the slit case to change abruptly from zero to a constant value; whereas, in the photographic pattern, the light strength varies gradually from a maximum to a minimum value. This sinuous variation in light amplitude causes more light to be diffracted into the two first-order directions than into the higher-order directions.

The photographic grating also differs from the simple slit grating in that, during the developing, fixing, and drying process of the photographic plate, the emulsion shrinks. Film, after exposure, still contains unexposed portions, and these cannot be allowed to be acted upon further by light, as the exposed portions were in the original exposure. The process of developing and fixing exposed film is designed to remove those light-sensitive elements still remaining unexposed in the emulsion, and the process causes the emulsion to shrink. The amount of this shrinkage, though quite small, is signifi-

cant for light waves, and it can cause a hologram to exhibit
certain *refraction* effects.

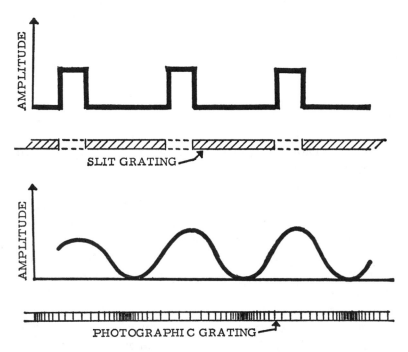

Figure 25. The classical slit grating (top) has abrupt changes in
light intensity, whereas the photographic grating (below) does not.

Refraction by Prisms

We recall that refraction is the process by which the direction of travel of wave energy is altered not by passing around
opaque objects (diffraction), but by passing through transparent objects, such as prisms or lenses. Waves, upon entering a refracting substance, experience a change in their velocity
of propagation. The simplest case of refraction occurs when
plane light waves enter a refracting substance whose surface
is tilted relative to their direction of travel. This case is shown
in Figure 26; light waves at the left, traveling in air, enter a

tilted block of glass. As they enter, they travel more slowly, and this lower velocity gives them the shorter wavelength shown—a result of equation (1) in Chapter 1. If the piece of glass is formed into a triangular prism, the light waves will emerge from the glass with their direction altered.

Plate 17 is a photo of a beam of sound waves being deflected downward by an acoustic prism. Here, the left side of

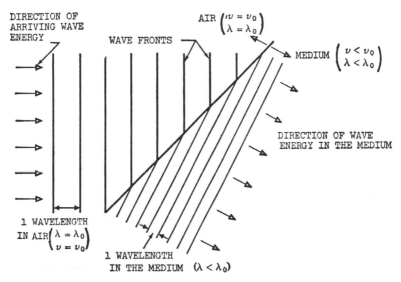

Figure 26. Wave energy entering a low-velocity medium experiences a shortening of its wavelength; the direction of propagation is altered accordingly.

the prism is parallel to the plane of the oncoming wave fronts; accordingly, no change in direction occurs as the waves enter the prism. However, because the right surface of the prism is tilted, the wave fronts are tilted downward as they emerge. This wave front tilt is portrayed in *Plate* 18.

A Prism Grating

Because emulsion shrinkage causes photographically made gratings to have varying thicknesses, similar prism or wave-tilting effects can occur for them. Consider a grating comprising, instead of many slits, many prism sections, as sketched in Figure 27. Each prism section tilts the wave direction downward as does the single prism of *Plate* 17. If, in addition, the light wavelength and the vertical prism spacing are properly selected, the light diffracted into the first-order downward wave direction can be greatly enhanced by the prism. If the spacing and wavelength are chosen so that waves moving in the direction of the prism tilt are also one wavelength behind (or ahead) of the waves issuing from their neighboring prism,

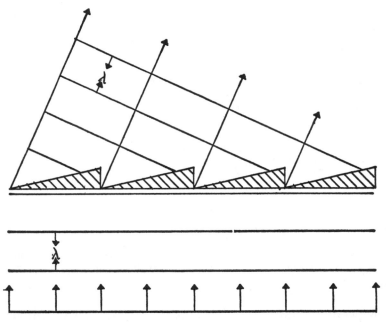

Figure 27. An array of prisms can act as a grating with the waves refracted by the prisms reinforcing the grating waves diffracted in a particular direction.

the two effects add. The situation would be similar to that of replacing each of the two sound sources in *Plate* 16 with prism radiators of the type pictured in *Plate* 17. If, in that arrangement, the downward-refracted prism lobe of Figure 27 would coincide with the first-order downward-diffracted lobe of *Plate* 16, a significant enhancement of energy would occur in that one direction.

Similar diffraction-refraction enhancement procedures have been employed in the design of some microwave lens antennas. There, the process is called *stepping,* or *zoning.* This procedure can be used in microwave applications, because the wavelengths employed often cover only a rather narrow band of frequencies, and they can, therefore, be treated as single-wavelength devices. A stepped or zoned circular microwave lens, used in the first microwave relaying by the Bell System of television programs between New York and Boston, is shown in *Plate* 19. *Plate* 20 shows two cylindrical refraction-diffraction lenses. In the top lens of *Plate* 20, there are twenty-four zones in each (horizontal) half; in the lower one, there are five in each (vertical) half. It is seen that each of the half-sections of these lenses somewhat resembles the cylindrical prism structure of Figure 27; they are different in that, because a focusing effect was desired, the spacing of the prisms is not uniform.

A Double-Prism Grating

An extension of the prism-grating concept of Figure 27 permits enhancement of both upward and downward first-order waves; this concept is sketched in Figure 28. Here double prisms are used, with the top halves acting as downward-tilting prisms, and the lower halves causing wave energy to be tilted upward. Again, it is assumed that the vertical spacing and the wavelength are correctly chosen to cause the prism tilt directions to correspond to the first-order diffraction directions.

If we assume that, after developing and fixing the photographically recorded grating shown at the bottom of Figure 25, some emulsion shrinkage has occurred, we see that the top and bottom horizontal lines of grating will become wavy lines. Ridges and troughs will have been created, with the troughs positioned at the areas of maximum shrinkage. We noted that even a small amount of shrinkage can be significant because of the short wavelength of light. Of importance, also, is the

fact that the spacing of these emulsion prisms is exactly right
to bring about the effects described in connection with Figure
28. The contour of the emulsion ridges (the "prisms") will
not be triangular, they will be wavy (sinuous); further, their
height (their crest to trough thickness) may not be exactly
correct for ideal prism action. However, even a small amount
of prism action can enhance the two first-order diffractions
relative to other order waves. The vertical extent or *aperture*
of each of the prisms is very small, perhaps only a few wavelengths; the beam of each of these tiny prisms is, therefore,
quite broad. (We note that the prism beam of *Plate* 17 is similarly broad.) Hence, because the proper prism *spacing* always

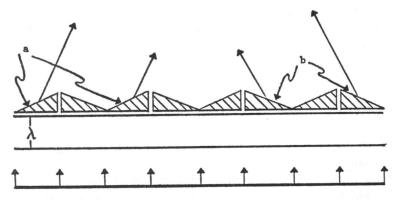

Figure 28. An array of double prisms, acting as a grating, can
cause both first-order diffracted waves to be reinforced. Emulsion
shrinkage in a photographically recorded grating can introduce
thickness effects comparable to this double prism grating.

exists (this is inherent in the photographic process), almost
any amount of emulsion shrinkage will cause some additional
wave energy to be tilted into the desired first-order direction.

Variation in the emulsion thickness caused by shrinkage
similarly affects the action of holograms. For some holograms,
the recorded black and white interference patterns are almost
completely eliminated through the process of bleaching the
photographic film. For these, the film plates become very
transparent; yet, they still exhibit the many surprising properties of ordinary unbleached holograms.

Volume Effects

A third difference between the simple slit grating and the photographic grating arises from the fact that the photographic emulsion is many light-wavelengths thick. The interference pattern generated by two sets of light waves is three-dimensional in extent. When this pattern is recorded photographically, it does not merely exist at the *surface* of the photographic film (as was tacitly assumed in connection with Figure 5); it establishes itself throughout the three-dimensional *volume* of the film emulsion. Because the film emulsion has an appreciable thickness, this volume pattern, as recorded in the photographic film, can be significant. Photographically made gratings, zone plates, and holograms must, therefore, be considered as recorded volume interference patterns. We shall discuss these volume effects in the next chapter.

Dependence of Diffraction on Wavelength

A grating diffracts differently colored light in different directions. This effect is sketched in Figure 29. Waves having the longer wavelength, λ_1, are one wavelength behind in the direction, A, whereas, those of wavelength λ_2 are one wavelength behind in a different direction, B. Because this property of gratings, that of diffracting wave energy of different wavelengths in different directions, is shared by holograms, let us analyze the effects it produces.

The diffraction pattern of *Plate* 16 was formed with single-frequency sound waves, and it is, therefore, analogous to the pattern formed with single-frequency (single color) light waves. Multicolor light waves behave differently. When the multicolors of white light are directed toward a slit grating, colored bands, that is, colored fringes, result. Similarly, if the two single-frequency sound sources in *Plate* 16 were caused to emit single-frequency sound waves having a longer wavelength than that of the sound waves used in the figure, the bright areas would appear in positions other than those shown. Similarly, in the pattern formed by a light wave grating, a wave of one color could generate on the screen a bright area of that color exactly where waves of another color would produce a dark area. A grating diffraction pattern for white

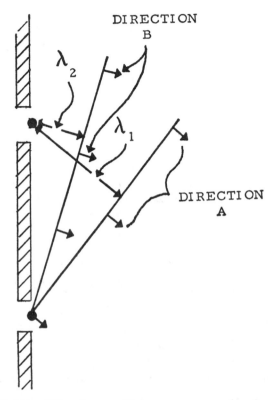

Figure 29. The diffraction angle for waves emerging from two slits or from a grating depends on the wavelength. Because holograms are a form of grating, their performance also depends upon wavelength.

light thus consists of many diffraction patterns, each one produced by each of the various colors. Because these various patterns are not registered on the screen, variously colored fringes result.

Diffraction by a Knife Edge

We turn now to the second kind of diffraction included in Webster's definition—diffraction caused by the edges of opaque bodies. When waves pass by such edges, some energy is de-

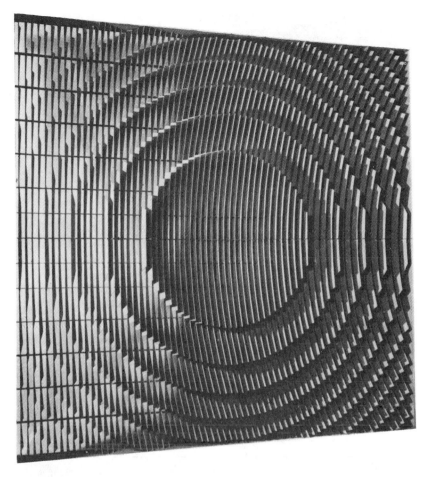

PLATE 19. The circular pattern of the zones in some microwave lenses resembles that of a zone plate.

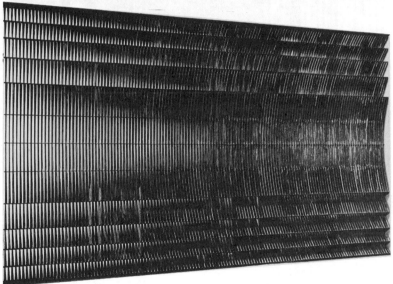

PLATE 20. These two cylindrical microwave lenses have zone structures which resemble prisms.

PLATE 21 *(opposite, top)*. Plane sound waves arriving from the left proceed unhindered at the top of the photo. In the shadow region, fainter circular wave fronts are evident, caused by diffraction at the edge of the shadowing object.

PLATE 22 *(opposite, bottom)*. The amplititude pattern in the shadow of a disk shows a bright, central lobe.

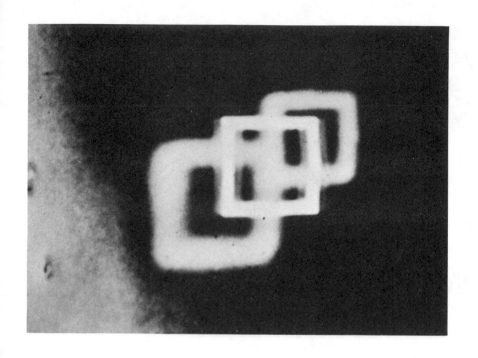

PLATE 23 *(opposite, top)*. A two-inch-diameter hologram zone plate forming three images. The zone plate is the dark area with a circular outline, and the object was a white square. The zero order (straight through) image is the bright square the virtual, diverging image is to its lower left; and the real converging image is to its upper right.

PLATE 24 *(opposite, bottom)*. One form of microwave antenna combines a horn with a portion of a parabolic-reflecting surface.

PLATE 25 *(above)*. A photograph taken with a camera using the two-inch zone plate of *Plate 23* as its only lens.

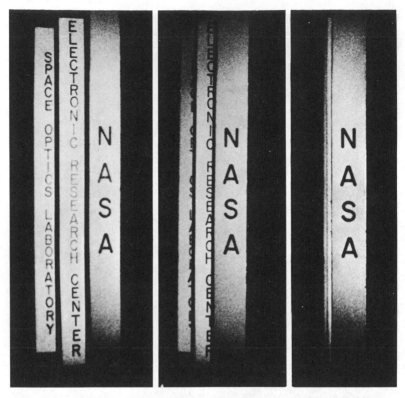

PLATE 26. Three photographs of a hologram being illuminated with laser light. The hologram recorded a scene comprising three vertical bars; for these photos the camera was moved successively farther to the right, finally causing the rear bars to be hidden by the front bars.

PLATE 27. Emmett Leith, the first to use lasers in holography, illuminates one of his holograms at NASA's Electronics Research Center. (Courtesy National Aeronautics and Space Administration)

PLATE 28. George Stroke, a pioneer in making laser holograms, first proposed the term "holography." (Courtesy University of Michigan News Service)

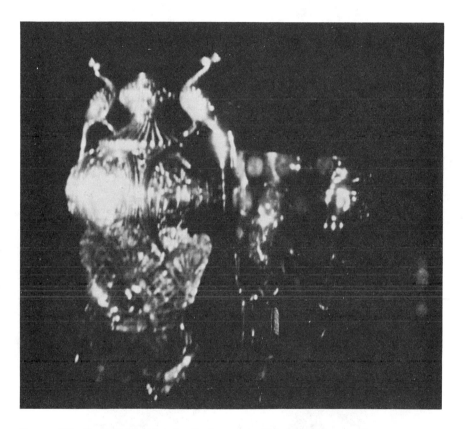

PLATE 29. A photograph of a hologram, showing the top of a cut-glass toothpick holder, a carved silver object, and a "thumb-print" glass. The viewer, as he moves his head, will see changes in the light reflected from these objects' surfaces.

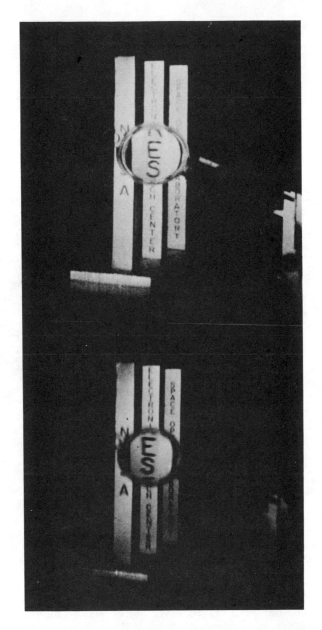

PLATE 30. A hologram of a lens shows the proper change in relative magnification when the observer's position, or the camera's position, is changed from (top) close viewing to (bottom) distant viewing.

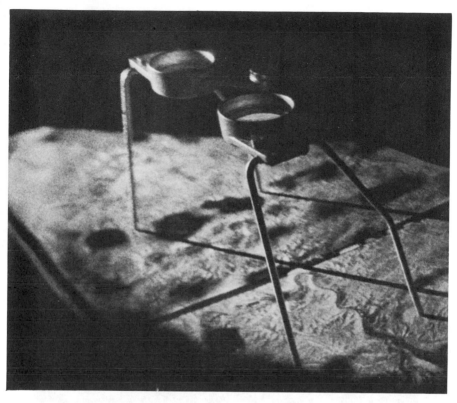

PLATE 31. A photograph of the objects used in making an unusual hologram. Two small lenses on a small stand act as stereo viewers for the plaster model of hilly terrain beneath them.

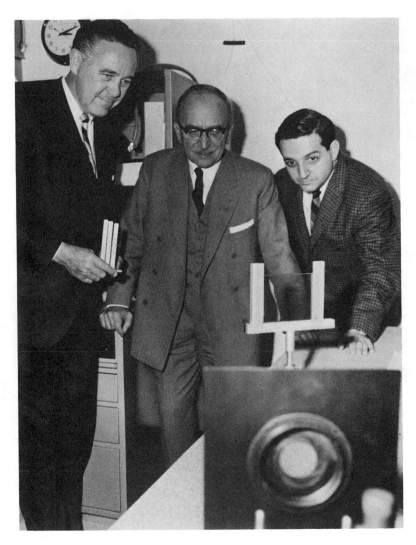

PLATE 32. NASA's Lowell Rosen (right) shows holography inventer Dennis Gabor (center) a reconstructed hologram while the author looks on. (Courtesy National Aeronautics and Space Administration)

PLATE 33. When the hologram used in *Plate 26* is illuminated with light from a mercury arc, numerous images appear, corresponding to the various colors (spectral lines) of the arc light.

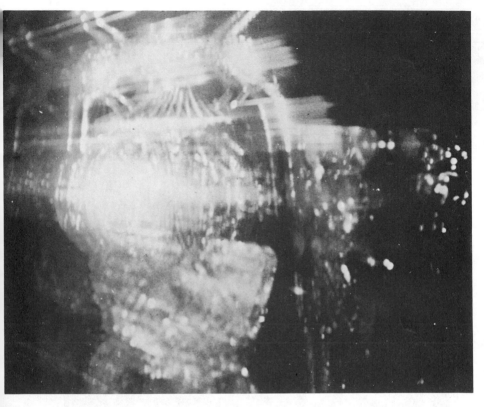

PLATE 34. Illuminating the hologram of *Plate 29* with a mercury arc lamp results in objectionable blurring.

flected (diffracted) into their shadow region. This phenomenon is illustrated in *Plate* 21. In this photograph, sound waves arrive from the left, and a long, wooden, rectangular board acts as a shadowing object. In the region above the board, the waves are seen to continue unimpeded toward the right as plane waves. Because the waves cannot penetrate the board (it is opaque to the sound waves), the lower right of the photo is a shadow region. The edge of the board, usually referred to as a "knife edge," is seen to act as a new source of sound energy. The faint waves in the shadow region have wave fronts which are circular (cylindrical), and the common axis of these cylinders is located at the knife edge.

Diffraction by a Disk

If the opaque object is, instead of a knife edge, a disk, energy is diffracted from all points on the perimeter of the disk, and a much more complicated pattern exists in its shadow. This wave pattern was portrayed in *Plate* 3 as an illustration of a complicated interference pattern generated when nonuniform sets of waves interfere. We can see, however, from the knife-edge diffraction of *Plate* 21 how sound diffracting around the disk edges can form this pattern. In *Plate* 3, sound waves also arrive from the left, and in the upper and lower parts of the photo, in the nonshadow areas, the waves proceed unhindered (as in *Plate* 21) as parallel wavefronts. In the shadow region itself, two sets of circular wavefronts are evident, one having the top edge of the disk as its center of curvature, the other having the lower edge as its center. Not fully discernable in *Plate* 3 are the circular wavefronts arriving from all points on the perimeter of the disk; these points all act as new sources of wave energy, and these many wave sets all interfere with one another. This complicated mixture establishes a narrow wave pattern along the axis, a pattern which looks much like the pattern of the unimpeded waves at the top and bottom of the photo. Thus, the combination of the many new wave sources positioned along the circular perimeter of the disk produces a concentration of wave energy along the axis.

Lord Rayleigh observed this effect for both light waves and

sound waves. Using a penny as his opaque disk and sunlight
as his source of light waves, he demonstrated the presence
of a bright spot in the center of the circular shadow of the
penny. From our everyday observations of the shadows cast
by opaque objects placed in the path of sunlight, we would
find it difficult to believe that the wave intensity or brightness
at a point in the deep shadow of a disk is, as Lord Rayleigh
stated in his *Theory of Sound,* "the same as if no obstacle
at all were interposed."

Rayleigh further described methods for demonstrating, with
acoustic waves, the existence of this bright region. He states
that the "region of no sensible shadow, though not confined
to a mathematical point on the axis, is of small dimensions"
and, further, that "immediately surrounding the central spot,
there is a ring of almost complete silence, and beyond that
again a moderate revival of effect." *Plate* 22 omits the wave
fronts of *Plate* 3 and portrays only the amplitude pattern of
sound intensity behind the disk (for this figure, the same disk
and same wavelength sound waves as those of *Plate* 3 were
used). In this figure, we see clearly all three of Rayleigh's
areas, the bright central spot or cone, the "ring of silence"
above and below the bright central cone, and the "revival
of effect" above and below the ring of silence. These same
areas can also be seen in the wave pattern photo of *Plate* 3.

Disks and Zone Plates

Rayleigh noted that the wave concentration in the shadow
of a disk can be further enhanced by positioning additional
annular rings around the disk, thereby blocking out other
areas or zones of wave energy. This process of adding additional blocking rings leads to a *zone plate,* a device used for
focusing several forms of wave energy. We shall see in the
next chapter that the optical zone plate is closely related to a
hologram because the photographic record of the interference
pattern of spherical waves and a set of plane reference waves
are forms of zone plates. A knowledge of the more pertinent
properties of zone plates can provide us with a better understanding of the hologram process.

Chapter 5
ZONE PLATES

In the description of holograms in the first chapter, it was noted that, by virtue of spherical light waves interfering with plane reference waves, each luminous point of a scene forms, in the recording process, its own circular interference pattern, its own zone plate, on the photographic plate. As noted earlier, the similarity between the photographic recording of interference patterns (holograms) and the optical diffraction device, the zone plate, was first recognized in 1950 by the British scientist, G. L. Rogers. Let us examine this relationship.

The Classical Zone Plate

The zone plate can be described as a set of flat, concentric, annular rings which diffract wave energy. The open spaces permit passage of waves which add constructively at a desired focal point, and the opaque rings prevent passage of waves which would interfere destructively with these waves at that point.

Figure 30 indicates the procedure for determining the positions of the rings in a zone plate having a circular opaque disk for its central portion. We recall from the previous chapter that an opaque disk can effect a concentration of energy along its axis by virtue of wave energy being diffracted by it into the shadow region. At some point along this axis, a further concentration of energy can be provided by the use of opaque rings which allow only that wave energy to pass by as will interfere constructively at that particular point on the axis.

The proper positions of the blocking rings are determined, as shown in Figure 30, by drawing circles whose centers coincide with the desired focal point and whose successive radii

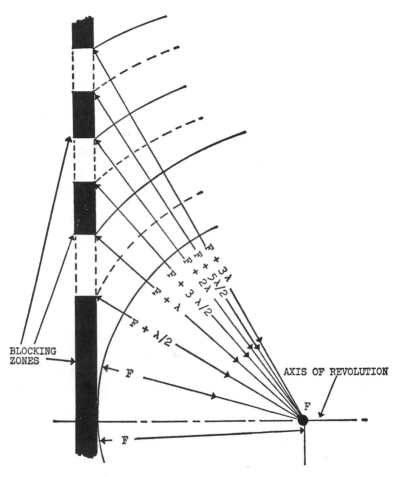

Figure 30. When the diffraction of a disk is supplemented by the added diffraction of annular rings, a zone plate results.

from this point differ from one another by one-half of the design wavelength. Thus, at that point where the first one-half wavelength circle intersects the plane of the zone plate, the central blocking zone—the opaque disk—is terminated. Farther out, at the one-wavelength point, the first annular blocking ring is started. It is terminated at the one and one-half wavelength point, and the process continues. It should be noted that the design of a zone plate is based on one particular wavelength, and that waves having wavelengths which differ from this design wavelength will not be affected by it in the desired way.

Zone plates can be used at light wavelengths; they can be used for concentrating microwaves; they can be used in acoustics. Lord Rayleigh noted that a zone plate designed for sound waves can play "the part of condensing lens." An antenna of a World War II microwave radar incorporated a zone plate, and some microwave lens antennas now in use employ the zone concept. In these uses, the microwave lens thickness is altered at the one-wavelength zone positions; examples were noted in the previous chapter. For focusing wave energy, zoned lenses are more efficient than zone plates because they avoid discarding the energy which is reflected or absorbed in a zone plate by the blocking zones.

Zone Plates as Negative Lenses

Lord Rayleigh compared his acoustic zone plate to a condensing lens, and the usual microwave and optical applications of zone plates similarly exploit only their focusing ability. The fact that zone plates also cause diverging waves to be formed is not too generally appreciated. In a hologram this negative lens property of a zone plate is very important; it is the hologram's diverging waves which give the viewer the striking, three-dimensional view of the original scene.

That zone plates do cause both converging and diverging waves to be generated is evident from design considerations. It is just as possible to design a zone plate which will cause waves to *diverge* as it is to design one which produces a converging effect. If one selects the point from which the waves are to diverge so that it is on the opposite side and at a distance from the zone plate equal to that of the focused point, one finds that these two designs are identical—the blocking zones are in exactly the same location for the diverging as for the converging zone plate.

It follows that when plane waves fall on any zone plate, both converging and diverging waves are generated. For optical zone plates, this simultaneous diverging and converging property results in multiple image effects; when a zone plate is placed in front of an object, several images are seen. This effect is shown in *Plate* 23. The zone plate used here was a photographic recording of the interference pattern formed by light waves from a point source interfering with plane waves. It is the darker circular area in the photo. The zone plate is imaging, off-axis, an object which looks like a square, white, picture frame. Of the three "images" seen, the object itself is the brightest and clearest because the zone plate is fairly transparent; this "image" is what we have called the zero-order component. On either side of it is another, slightly blurred image of the white frame. The one deflected toward the center of the dark circular area is produced by the converging waves, and the other is formed by the diverging waves. It is worthy of note that here only the zero-order and the two first-order images are prominent. This would not be the case if a classical zone plate were involved. The gradual change from transparency to opacity in the fringes of the photographic (hologram) zone plate causes the higher-order images to be suppressed.

Complete separation of images has not occurred in *Plate* 23 because the object is too near the axis of the zone plate. Had it been placed further off-axis, all three images would have been fully separated and would appear as separate, individual images.

Zone Plates with Areas Interchanged

One property of zone plates pertinent to holography is that their functioning is unaffected if the blocking zones and open-area zones are interchanged. This is evident from Figure 30. If the central and other blocking zones in that figure were to be made open-area zones, the wave sets passing through them would still differ from one another at the focal point by exactly an integral number of wavelengths., Additionally, since the originally open areas (now blocking areas) differ by a half wavelength from the new open areas, the desired zone plate action will be unaffected.

This ability of a zone plate to function properly with transparent and opaque zones interchanged can be even further extended; the center area, whether blocking or open, can be made any size at all. Then, it is necessary only to make the next zone area of such a size that the distance from the desired focal point to its outer edge is one-half wavelength greater than the length to its inner edge; the next area is one-half wavelength greater than that, etc. Again, full zone plate action results, since additive waves are passed and canceling waves are blocked.

It is this property of interchangeability of zone plates which permits them to be so easily recorded photographically. This can be seen from a consideration of the three-pinhole hologram of Figure 9, for which each pinhole is at a different distance from the plane of the photographic plate. For one of the three-pinhole interference patterns, the central area might be strongly exposed, causing a central, blocking zone to be formed. For the other two pinholes, the central zone might turn out to be a full open area, or a small open area, or a blocking area. Since zone plate action requires only that successive areas be positioned with half wavelength differences in their distance to the focal point, and since the circular wave patterns inherently have this property, all three photographic-recorded pinhole patterns of Figure 9 automatically become effective zone plates.

Offset Zone Plates

To clarify the action of an offset zone plate, let us review certain characteristics of two focusing devices, the parabola and the zone plate. Parabolic reflectors are often used as microwave antennas. The ability of such a reflector to form a beam of plane waves is traceable to the geometrical property of a parabola described in Figure 31. Very often, only

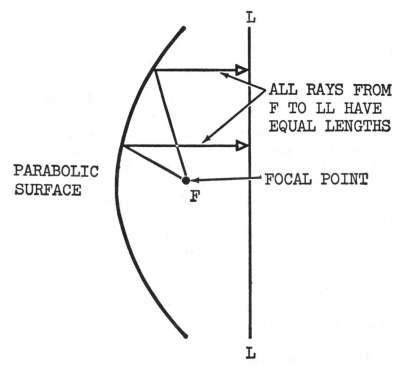

Figure 31. Spherical waves issuing from the focal point F are converted to plane waves by a paraboloidal surface because the times of travel to the plane LL are alike for all rays.

portions of such reflectors are used. The antenna shown in Figure 32 is such a device; only part of the parabolic (paraboloidal) surface is used to reflect incoming energy into the horn. It is an offset reflector, and one form of this antenna, now used widely in microwave relays, is shown in *Plate* 24.

Similarly, an offset section of a zone plate can focus wave energy. This procedure is sketched in Figure 33. Plane waves arriving from the left impinge on a portion of a zone plate, which causes energy to converge on the focal point, f. In

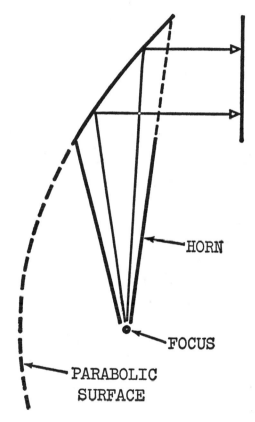

Figure 32. Part of a parabolic reflector can be used to produce flat wave fronts.

addition, this portion will, through its negative lens action, generate diverging waves appearing to emanate from the conjugate focal point, fc.

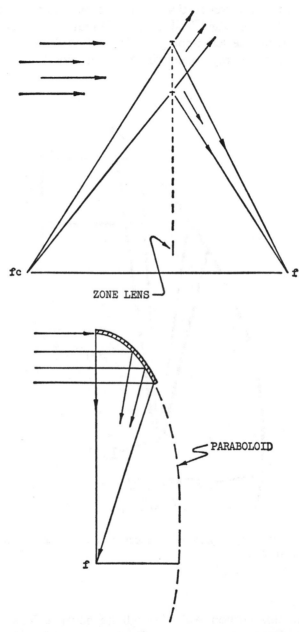

ZONE LENS

PARABOLOID

Figure 33. Portions of a parabola (right) or a zone plate (left) can
focus (and diverge) wave energy.

Offset Holograms

In making the first laser holograms, Leith used a prism to tilt the reference wave and cause it to interfere with the waves issuing from a photographic transparency. Holograms made with this prism technique of Leith have been referred to as two-beam or offset holograms.

The offset terminology is used to distinguish such holograms from Gabor's original ones in which a single light beam was used; there, the light passing around the object acted as the reference wave. Actually, the single-beam technique also can be used to form an offset hologram, as was shown in Figure 9. There the single beam of laser light falls on the pinhole screen and also passes above it. As noted, when the plate is developed and illuminated with laser light, both real and virtual images of the original scene (the point source of light) are formed.

Zone Plates as Lenses

Photographically made zone plates have been suggested for use, as lightweight lenses, in the fields of communication, astronomy, and space. They could be photographically recorded on lightweight plastic sheets occupying very small space and later "unfurled" to their full aperture size for concentrating laser communication beams or for taking photographs. That photographically made zone plates can act as optical lenses is shown in *Plate* 25. This photo was made with a "camera" having the hologram zone plate, used in making *Plate* 23, as its only lens.

Volume Zone Plates

Zone plates used for microwaves and sound waves, and the ruled glass zone plates used for light wavelengths, are generally considered as being two-dimensional (planar) structures. Because of the extremely short wavelengths of light waves, photographically made zone plates exhibit three-dimensional or volume effects, and this effect must be taken into account. The emulsion in a photographic plate may be twenty light wavelengths thick, so that three-dimensional interference patterns become recorded in the emulsion.

The use of such three-dimensional interference patterns was first employed in holography by George Stroke and his co-workers to obtain white light or reflection holograms. Photographically made zone plates are actually a form of such volume holograms, and their three-dimensional characteristics cause their performance to differ from that of simple two-dimensional zone plates, just as the volume effects in reflection holograms cause their performance to differ from that of conventional holograms. These photographically recorded volume effects for zone plates also provide the basis for reflection holograms and three-color holograms.

Standing Wave Patterns

As noted in the previous chapter, interference effects exist not only at the surface of the photographic plate, but throughout the volume of the emulsion so .that if the photographic emulsion is many wavelengths thick, it will be exposed in depth and will record a volume interference pattern. The possibility of recording three-dimensional (volume) light wave patterns within the emulsion of a photographic film was recognized in the early 1900s by the French scientist, Gabriel Lippmann. He proposed using such recorded patterns for a form of color photography. His patterns were *standing wave* patterns, and because these are also of importance in certain holograms and zone plates, let us review the characteristics of standing waves.

In the resonator diagram of Figure 23, various resonances were indicated, having wavelike outlines or envelopes. The outlines are formed by virtue of the waves bouncing back and forth between the two end walls, but because the pattern these moving waves create is stationary, the waves are called standing waves. This standing wave pattern can also be looked upon as being made up of two waves, one moving to the left, the other moving to the right. Standing waves thus can be considered as being formed either by a reflection of wave motion at a wall or by two single-frequency waves moving in opposite directions. It is this latter, oppositely moving wave case that is of interest to us in holography and zone plates.

In Figure 34, plane wave laser light is shown arriving from the top of the figure, and it is split into two horizontal beams

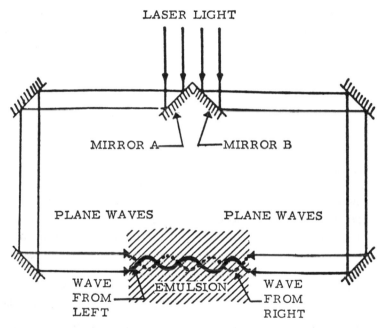

Figure 34. Coherent waves traveling in opposite directions form standing waves which can be recorded in a photographic emulsion as densely and lightly exposed planes.

by mirrors A and B. By means of the other four mirrors, the two beams are caused to be superimposed and to interfere with each other while traveling in opposite directions. They thereby generate a standing wave pattern. In the figure, this is shown as being formed within the emulsion of a photographic plate. It is evident that there will be regions (planes) where destructive interference (low light intensity) exist; in these areas, the emulsion will be very lightly exposed. At the in-between, *strong* light intensity planes, the emulsion will become more heavily exposed. When the emulsion is developed and fixed, it will exhibit heavily exposed vertical planes spaced apart at one-half wavelengths of the original laser light.

Lattice Reflectors

The half wavelength lattice structure just described is known to be highly reflective for waves whose half wavelengths correspond to the lattice separation. This reflection phenomenon can be explained with the aid of Figure 35. Here waves arriving from the right are reflected successively by the numerous, rather densely exposed (but not opaque) planes 1, 2, 3, and 4 located within the photographic emulsion. Some waves, A, are reflected at the first surface, 1. Other waves of energy which pass through the various surfaces are indicated as waves B, C, and D. Waves B, when reflected at plane 2, rejoin waves A at a distance, 2a, behind them. Similarly, waves C are a distance, 4a, behind them, and waves D, a distance, 6a. If a is equal to one-half wavelength of the incident light, all three of these additional distances are exact multiples of one wavelength, and all reflected wave sets, A, B, C, and D, will be in perfect step; the crests of all will coincide as will the troughs. Thus, for light of wavelength 2a, reflections from all planes will add constructively, and the *reflectivity* of this structure to such wavelength light will be very high.

Because other wavelength light will not add in this way, the reflectivity for them will be significantly lower. Accordingly, when this multiple structure is illuminated with white light, which comprises all the colors (wavelengths) of visible light, it will reflect only a single wavelength (single color) light. If, in the developing and fixing of the photographic plate, no shrinkage has been permitted (which is not easy to accomplish), the color of this reflected light will be the same as that of the laser light which was responsible for forming it.

This procedure is the basis of the Lippmann color process;

as noted, it was recently used to permit holograms to be viewed with white light instead of laser light (these are called reflection holograms). The process was later extended to making *color* reflection holograms, exposing the hologram plate, successively, with two or three differently colored laser wave sets.

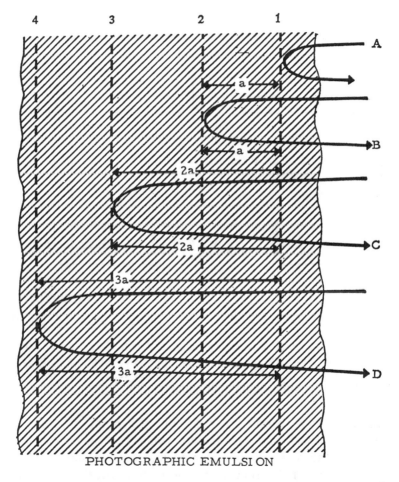

Figure 35. Reflecting planes, recorded in the photographic emulsion of Figure 34, cause waves having a wavelength equal to twice the plane spacing, a, to be reflected as constructively interfering waves.

Reflection Zone Plates

Zone plates can also be made using this three-color reflection procedure. Plane and spherical waves are made to approach the photographic plate from opposite sides, as shown in Figure 36. Their interference generates a zoned longitudinal

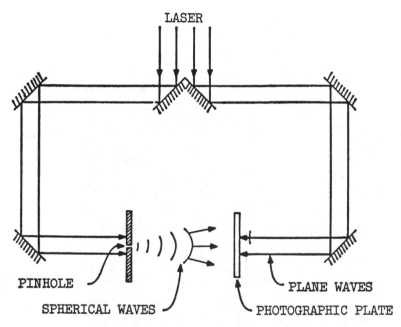

Figure 36. By causing coherent plane and spherical waves to impinge on a photographic plate from opposite sides, a reflection zone plate is produced. The process is similar to that used in forming reflection (white light) holograms.

lattice structure within the emulsion. When the emulsion is developed, fixed, and illuminated with white light, reflection and focusing of only one color occurs; the remaining colors pass through. If this reflection zone plate is used as a lens or focusing device, a monochromatic image is formed.

To make a *three*-color zone plate (a zone plate camera lens), the photographic plate would be exposed, as shown in

Figure 36, not once, but three times, using three differently colored laser sources, successively. The developed emulsion would then contain three longitudinal lattices, which, when exposed to white light, would reflect and focus only those colors of the white light having wavelengths corresponding to the three different lattice spacings. If this reflection lens were to be used as the lens in a camera employing standard color film, it would properly image three differently colored views of the scene on the film, and satisfactory color reproduction of objects or scenes would result.

In the last chapter, we noted several differences between ordinary gratings and photographically formed gratings; these differences also exist between ordinary and photographic zone plates. In the classical zone plate, the zones are either completely opaque or completely transparent, whereas a photographically made (hologram) zone plate has zones whose densities vary gradually from their most opaque values to their most transparent values. This property of holograms and photographic zone plates causes their first-order diffracted waves to be accentuated, as was noted by Rogers. The existence of shrinkage in the emulsion results in a varying thickness photographic zone plate; this also accents the first-order diffraction. Also, hologram zone plates differ from classical zone plates because the thickness of the emulsion is equal to many light wavelengths, so that the classical two-dimensional (planar) zone plate structure is replaced by a three-dimensional one. These same properties of photographic zone plates are also exhibited by holograms.

Chapter 6

PROPERTIES OF HOLOGRAMS

In this chapter, the unusual features of holograms will be reviewed. Several of these were considered so surprising when first described in the literature that numerous expressions of disbelief were voiced.

Three-Dimensional Realism

The most spectacular characteristic of holograms and the one most surely responsible for the tremendous interest they have evoked is their ability to create an extremely realistic illusion of three dimensions. This property is difficult to demonstrate, lacking an actual hologram with its needed illumination. However, some idea of the effect can be gotten by looking at several different photographs of the same, properly illuminated, hologram. Three such photos are shown in *Plate* 26. The scene recorded in this hologram comprised three vertical bars, placed one behind the other, each having letters placed vertically along the bars. It was made, as described, for the hypothetical sphere and pyramid hologram of Figure 3. The three bars become clearly visible when the hologram is illuminated with laser light. From certain angles, the observer can see all three bars; however, from one particular direction, his view of the rear bars is blocked by the front one. The camera was placed in this latter position for the photograph shown at the right of *Plate* 26. For the center photo, the camera was moved slightly to the left, partially exposing the two rear bars. For the left-hand photo, it was moved still further to the left, fully exposing all three bars.

In viewing an actual hologram, the three-dimensional illusion is far more realistic than these three photos can convey. The viewer quickly realizes that much more information about the scene is furnished by a hologram than by other three-dimensional photo processes, such as by stereophotography, using a pair of stereophotos. In the hologram reconstruction, the viewer can inspect the three-dimensional scene not just from one direction, as in stereophotography, but from many directions. Because this property of complete realism has undoubtedly given the development of holography great impetus, the chronology of three dimensions in holograms may be worthy of brief discussion.

As early as 1949, Gabor wrote, "the photography contains the total information required for constructing the object, which can be two-dimensional or three-dimensional." However, he used, for his objects, two-dimensional transparencies, and said little as to how the recording and reconstruction of three-dimensional objects would be accomplished. Accordingly, in December 1963, when Leith, who was an esteemed contributor in the radar field, first used lasers in holography, he likewise employed two-dimensional transparencies for his objects.

Leith thus pioneered in producing the first laser holograms and the first offset holograms, but his first objects and images were still two-dimensional. Also, he discussed his early hologram experiments in terms of the optical processing technology of coherent radar; we shall note later the close relationship between holography and this form of radar. Later, Leith and his group extended their concepts to three-dimensional reflection holograms. *Plate* 27, taken at the NASA Electronics Research Center when the author was its Director, is a photo of him and one of his very spectacular holograms being reconstructed by the laser beam at the left.

George Stroke (*Plate* 28), a recognized authority on the ruling of precision optical gratings, started quite early to look upon holograms (as has this book) as diffraction devices. In a May 1964 set of lecture notes at the University of Michigan, he described how light waves, reflected onto a photographic plate by two adjacent, slightly tilted plane mirrors, generate a photographic grating, and how, when a three-dimensional object is substituted for one of the two mirrors, a three-dimensional hologram results. It was not easy for some to accept this generalization to three dimensions, which Leith and Stroke described. Thus, in the discussion period following a lecture by Stroke on holography in Rome in September 1964, an eminent Italian scientist remonstrated, "The light beam cannot carry information about a three-dimensional object because this is described by three degrees of freedom, whereas a light beam has only two degrees of freedom." Although the logic of this objection appears, at first sight, quite reasonable, we can assure ourselves, for example, by a consideration of Figure 10, that three-dimensional

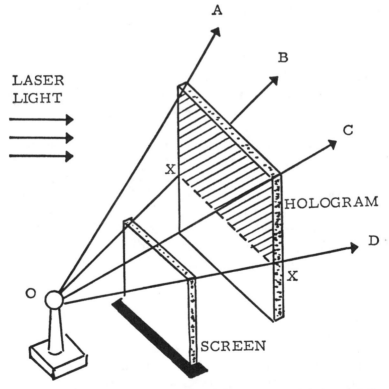

Figure 37. A hologram of a point light source which is partially hidden from the hologram by an opaque screen presents a paradox —on reconstruction, can the ethereal, non-existent screen still hide the imaginary light source?

information *can* be recorded on a two-dimensional surface by means of the hologram (zone plate) process.

The three-dimensional realism of certain holograms is often difficult for some to understand. Thus, Figure 37 shows a hologram recording a scene which includes a small, very bright reflecting point having an opaque screen in front of it. In the actual scene, a viewer positioning himself to be below the shadow demarcation line, A–A, would have his view of this point source quite understandably blocked by the opaque screen. In the reconstruction, however, says the skeptic, the screen is actually not there; it is a ghostlike, evanescent figment of the imagination, so how can it possibly block the reconstructed light issuing from the point?

Here again, the zone plate analogy provides the answer. In the recording process, only a portion of the reflection point's zone plate is recorded on the photographic plate (it is shown as a shaded area). The remainder (below line A–A), having been blocked by the screen, is missing. Accordingly, as far as the luminous point itself is concerned, its hologram "window" ends at the line, A–A, and its (partial) zone plate can only diffract laser light in directions encompassed by the pyramid, OABCD. The point light source *can* be blocked by an apparition!

Holograms and Photographs

There are two ways in which the hologram photograph differs from an ordinary photograph. When a photo is taken of a scene, a negative is first made, then the dark and light areas of this negative are reversed to form a positive print which portrays the scene in its original form. In holography, either the positive or the negative version of the hologram generates the identical three-dimensional illusion. This property follows from the similarity between holograms and zone plates; we noted that zone plates and gratings perform equally well if their dark (blocking) and their bright (transparent) areas are interchanged. This process is equivalent to changing a positive print to a negative one or vice versa; it constitutes a major change in photography but no change at all in holography.

The second way in which holograms differ from photographs is in the appearance of the photographic plate. When a hologram is held up to the light, scarcely any pattern is seen, certainly none which is indicative of the scene recorded, as in a photographic negative. The hologram appears to be a uniform gray sheet, and it reveals none of the characteristics or features of the scene recorded until it is properly illuminated.

Parallax and Lens Action

In viewing a hologram, the observer is usually encouraged to move his head sideways or up and down so that he may grasp its full realism by observing an effect called *parallax*. In a real scene, more distant objects appear to move with the viewer, whereas closer objects do not. Such parallax effects are very noticeable to a person riding in a train; the nearby telephone poles move past rapidly, but the distant mountains appear to move with the traveler. Similarly, the parallax property of holograms constitutes one of their most convincing proofs of realism.

Because hologram viewers invariably do move their heads to experience this parallax effect, ingenious hologram designers often include cut glass objects in the scene to be photographed. In the real situation, glints of light are reflected from the cut glass, and these glints appear and disappear as the viewer moves his head. This effect also occurs for the hologram, and further heightens the realism. *Plate* 29 is a photograph of a hologram, comprising, in the left foreground, the top of a small cut glass toothpick holder; in the left background, an extensively carved silver object; and on the right, an early American "thumbprint" glass. All three of these objects show marked changes in the light reflected from various areas as the viewer moves his head, and the hologram similarly manifests these light variations.

The silver chalice used in this hologram was acquired by

the author in 1936 in Bangalore, India, during postdoctoral study with Sir C. V. Raman. Also visiting Raman in 1936 was Max Born, recipient of the Nobel Prize in Physics in 1954 and author of the ageless classic, "Optik." (An English edition of that book is now available: Born and Wolf, "Principles of Optics.")

Another proof of the realism of holograms is provided when a lens is included in the recorded scene. Then, instead of a sideways motion of the viewer, a motion toward and away from the hologram gives him the expected enlarging and contracting of objects behind the lens. This phenomenon is shown in *Plate* 30, which portrays two different photographs of the same hologram. The scene recorded contains a reading glass placed in front of some lettering. The top photo was taken with the camera close to the hologram; for this position, the magnifying glass includes the letters, E and S, and a portion of the upper letter, R. In the lower photo, the camera was moved some distance back, and the change in magnification is evident. The effect is heightened with increased distance.

A Stereo Hologram

A particularly interesting hologram involving lenses was made at the Bendix Research Laboratories and exhibited in March 1966. A sideways view of the scene involved is shown in *Plate* 31. The objects consisted of a pair of stereo viewing lenses positioned to permit a viewer, placing his eyes in close proximity to the stereo lenses, to observe a three-dimensional (stereo) view of a model of some mountainous terrain. The hologram was made with the hologram plate in very close proximity to the lenses. To view the reconstructed scene, the observer positions his eyes as though the two viewing lenses are actually in place. When he looks directly into the imaginary lenses, he imagines he sees a magnified view of the terrain in three-dimensional stereo, just as he would through the original stereo viewing lenses.

Focused-Image Holography

Another interesting use of lenses in holograms was made in 1966 by NASA scientist Lowell Rosen in the process now called "focused-image" holography. The procedure is shown in Figure 38. A lens forms a real image of a group of objects located behind the lens, and a hologram is then made of this (upside down) real image. The photographic plate can even

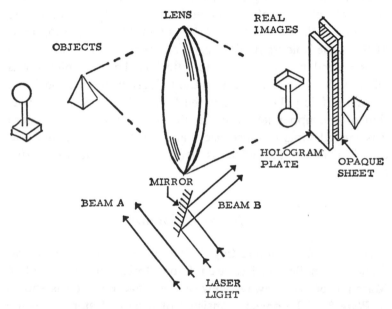

Figure 38. The images at the top left are focused by the lens and their real images recorded as a hologram. Even with the opaque screen (at the far right) present (so that one image is not formed), the hologram records both images. On reconstruction (with the opaque screen removed) both are presented to a viewer, the originally non-existent one standing out in space in front of the hologram.

straddle objects separated longitudinally in the real lens image; that is, the image of one object can be positioned behind the photographic plate and the image of a closer object positioned in front of the hologram plate. When this hologram is developed and properly illuminated, the viewer sees one object behind the hologram plane and another standing out in space in front of the hologram.

When the focused-image hologram process was first described, it was assumed by some that the presence of a lens would automatically cause a collapsing of the image onto the photographic plate. A joint publication on this subject by Stroke, Rosen, and the author noted that it had generally, "been accepted without question in the past that the 'third dimension' in a conventionally-focused photographic system would be irretrievably lost in the recording on simple, two-dimensional photographic plates or film." It continued: "In the normal photographic process, the camera lens forms real images of the objects of a three-dimensional scene, and these images exist in three-dimensional space; however, the ordinary photographic film causes these images to be collapsed into a single plane."

We noted that if the plate in Figure 38 simply had been exposed to the light from the objects (without the simultaneous use of the reference beam, B), the result would have been an ordinary photograph, and the images *would* have been collapsed in the resulting picture. The hologram process, however, retains the longitudinal, three-dimensional image separation.

It is of interest to consider what would happen if, as shown in Figure 38, an opaque sheet were to be positioned, so that it, too, straddles the two real images. This sheet, being opaque, would prevent the pyramid image from forming, and a viewer who could see both images if plate and sheet were removed, would see neither. A hologram, however, always records the wave pattern existing at its plane, and this original wave pattern possesses the intrinsic ability to form the pyramid image (as it would if the opaque sheet were removed). Accordingly, the presence of the opaque sheet has no effect on the hologram recording, nor on its later ability to establish real images of both objects in the reconstruction process. The hologram has recorded an image, which, with the opaque screen in place, does not exist, and it can later establish this "nonexistent" image in its reconstruction process (with the screen then removed, of course).

Interesting effects are obtained in focused-image holograms when an actual object is placed adjacent to the upside-down real images. For it to be recorded, it must, of course, be *behind* the hologram plate, a restriction which, we just saw, does not apply to the real images. The parallax effects are not the same for this true object and the imaged objects, and a most intriguing and confusing scene results when the hologram of such a combination is viewed. At about the time that Lowell Rosen was first experimenting with focused-image effects, Dennis Gabor visited the NASA center in Cambridge; a photo taken there of Rosen, Gabor, and the author is shown in *Plate* 32.

Reconstruction with a Small Portion of a Hologram

We observed in connection with Figure 33 that focusing devices such as paraboloidal surfaces or zone plates can achieve a focusing effect even if only a portion of their focusing surface is utilized. Because a hologram is a form of zone plate, it, too, can achieve certain of its effects if only a small portion of the original hologram is used in the reconstruction process. Figure 39 portrays this property for a hologram zone plate; both the full area, ABCD, and its smaller portion, A′ B′ C′ D′, generate real and virtual images.

Let us consider what effect this size reduction has on the zone plate of Figure 39. Because less light falls on the smaller area, A′ B′ C′ D′, less light is diffracted toward the focal point, f, and the focused, real image is, therefore, not as bright. In a similar way, the real image of a full *hologram* is made less intense, less bright, by a reduction in the hologram size. The original scene is still imaged however, and it can be portrayed on a white card located at the focal plane.

The size reduction also lowers the light intensity in the case of the virtual image of a hologram, but in addition, the realism of the image is reduced. Thus, in Figure 39, an observer viewing the virtual image of the point source of light will be required, for the smaller area, A′ B′ C′ D′, to position him-self more carefully in order to see the source through this new, smaller window, and this smaller window reduces ap-preciably the realism he would obtain in a full hologram through the parallax effect. Even though, with the smaller window, the viewer can still move his head about and see all of the objects in the scene, the window area he is using does not move with him, and he merely sees small, individual, al-

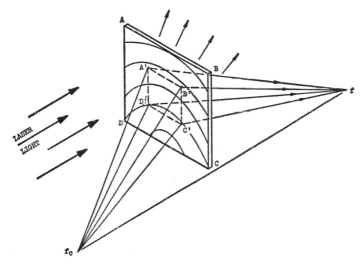

Figure 39. A portion A′ B′ C′ D′ of an offset zone plate ABCD diffracts converging energy toward f and diverging energy which appears to originate at fc. A portion of a hologram is similarly able to generate its real and virtual images, with certain limitations.

most two-dimensional views. As in the case of a true glass window, the viewer's ability to see around objects in the fore-ground is limited when the window size is reduced to a signifi-cantly smaller area.

In this connection, when the complete hologram is avail-able but is illuminated only by a small area "spot beam" of laser light, interesting things happen. As this small-area beam of light is moved about over the hologram area, the viewer of the virtual image can, by moving his head about, see all views, from all angles, of the recorded scene. A white card placed at the focal area will similarly display the real image on it of a series of changing, two-dimensional views of the scene. This latter ability to change, for an observer, the direction from which he views the original scene could have interesting applications. An aircraft pilot, attempting to land his plane in fog could have presented to him a prerecorded hologram picture of the airport runway from the very angle —up, down, or sideways—that he would be observing if the weather were clear.

Pseudoscopy in the Real Image of a Hologram

In the discussions of hologram zone plates, it was noted that they can function as converging lenses, so that the real image they form can be recorded on film just as is the real image of a camera lens. A hologram of a scene also produces a real image of the scene, and this image can be viewed on a white card placed at the focal area, just as the real image of a camera lens can be focused on a ground glass screen. When the scene recorded by the hologram is a two-dimensional transparency, the real image it reconstructs is also two-dimensional, and it is then indistinguishable from the real image a lens would form of that transparency. When, however, the original scene is three-dimensional, the real image of a hologram is also three-dimensional, but it then differs significantly from the real image which a lens would form of the same three-dimensional scene.

This effect is shown in Figure 40. The viewer is assumed to be at the right of the figure. As an object is moved further to the left of the focal point of the lens, as from B to A, the position of its image likewise moves to the left, as from B' to A'. Objects A' and B' in this lens-produced real image thus bear the same relationship to the viewer as do the actual objects, A and B.

For the hologram real image, this is not the case. In the recording process, luminous object A forms its own zone plate, and the focal point of this, when developed and illuminated, is, therefore, at the equidistant point, A'. Similarly, object B forms its own zone plate, with the focal point at its equidistant point, B'. For the viewer of this real image, objects originally in the rear appear to be in the foreground and vice versa. This reversal of forward-and-back objects has been given the name *pseudoscopy;* the image is said to be pseudoscopic. Because such images are rather confusing, little attempt has been made to utilize the real image for viewing purposes. However, scientists at K.M.S. Industries in Ann Arbor, Michigan, used a circularly symmetrical object (a cham-

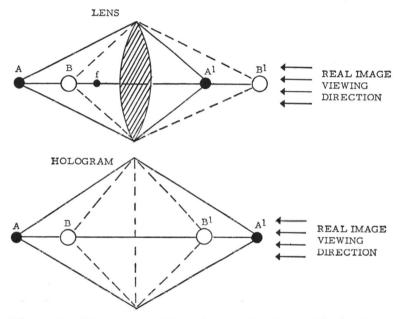

Figure 40. Objects imaged by a lens retain their original relative positions. When a hologram is made, each object forms its own zone plate, and, on reconstruction, the real and virtual images are positioned equidistant from the hologram; image inversion (pseudoscopy) thereby occurs.

pagne glass) as a hologram subject. In the reconstruction process, they presented the viewer with the real image, which stood out in front of the hologram as sketched in Figure 41. Because of the symmetry of the champagne glass, the exchange of front for rear was not significant, and the viewer imagined he was seeing a normal, nonpseudoscopic image.

Others had earlier proceeded along different paths to obtain "out-in-space" images. We have already mentioned one process; that of focused-image holography. Another procedure used was to make a second hologram of the pseudo-scopic image of a first hologram. The real image of this second hologram then becomes a "pseudoscopic-pseudoscopic" image

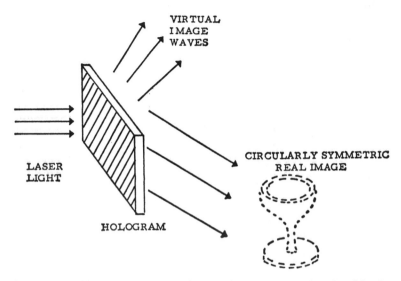

Figure 41. Notwithstanding the pseudoscopy associated with the real image of a hologram, such an image can be made to stand out in front of the hologram very realistically, provided it is circularly symmetric.

(a doubly reversed image), that is, a normal image. In this process, any scene, including unsymmetrical ones, can be used as subjects.

Image Inversion

As suggested in the sketch of Figure 41, both the real and virtual images can be presented to the viewer. Usually, as in Figure 41, they are widely different angles of view, and the viewer cannot see both simultaneously. He must either reposition himself to see the second image, or he must reposition the hologram.

Figure 42 shows how a simple rotation of a hologram can often accomplish this. When this (point source) hologram is rotated a half turn as shown, a viewer placed so that he

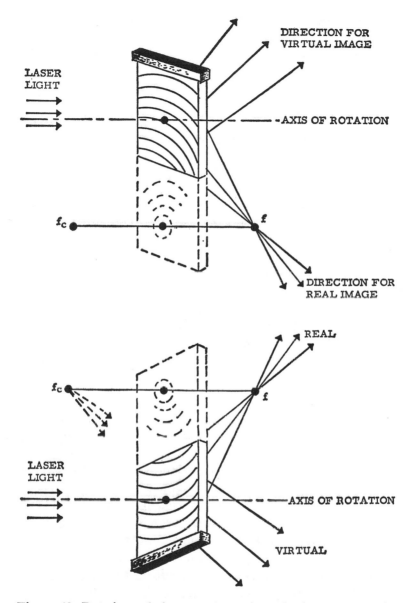

Figure 42. Rotating a hologram about the axis shown causes the
viewer to switch from the real image to the virtual image. Since the
real image was originally inverted, it now is upright; hence, unlike
a photographic transparency, turning this hologram upside down
does not turn the viewed image upside down.

originally saw one of the images now would see the other. A hologram made in this way of an actual scene would also behave in this manner. Because the viewer is usually at some distance behind the focal point, f, the real image, having passed through the focal point, is inverted, as for any focusing lens. Were he to have changed his own position to see the other image, he would have observed that the two images are inverted relative to one another. In switching his view from one to the other by a rotation of the hologram, however, the second image is rotated top to bottom. Thus, if the image he viewed first was right side up, the new image would also be right side up, even though he has turned the hologram upside down. Rotations about other axes also cause curious effects.

When in 1965, the author published a note on this phenomenon, a wide response again resulted. A former colleague sent a letter saying, "Beware! Other people's holograms don't work that way"; an M.I.T. professor explained the phenomenon very elegantly with Fourier transforms; and a third correspondent wrote: "The . . . properties . . . are exactly what one should expect."

Single Wavelength Nature of Holograms

One of the properties of ordinary holograms (and zone plates) which limit their usefulness is their single-frequency nature. We saw earlier that the design of a zone plate is postulated on one particular wavelength, and that for a given zone plate, only waves of one wavelength will be properly focused. Inasmuch as holograms are a form of zone plate, they, too, suffer from this problem; only single-frequency light waves can properly reconstruct their recorded images. If light comprising many colors is used in the reconstruction process, the various colors are diffracted in different directions, and the picture becomes badly blurred.

The photos of *Plates* 26 and 29 were made using single-frequency laser light for illuminating the holograms. For *Plates* 33 and 34, these same holograms were illuminated with light from a mercury arc lamp. This light source consists of many single individual colors, and the hologram causes each color to be diffracted in a different direction. Very objectionable blurring thus results.

This effect in holograms, whereby various colors are diffracted differently, makes it difficult to form a three-color hologram simply by exposing the photographic plate to three different-colored laser illuminations. In reconstruction, the illumination by three laser beams of this "compound" hologram would, in general, create nine images, only three of which would be in proper registry. The technique employed in making reflection holograms is one way to overcome this difficulty. The procedure is identical to that described in the last chapter for making three-color reflection zone plates.

Requirements on Film Properties

Light wavelengths are extremely short, and the recording, on film, of a light wave interference pattern demands film which has a very high resolution capability. All photographic films have a coating or emulsion placed on a glass or flexible film base. The emulsion is a mixture which usually includes a form of silver salt as its light sensitive ingredient. This mixture of materials can be finely or coarsely divided, and the fineness of the grains of the light-sensitive ingredient determines how much detail can be recorded photographically. When a recorded film is enlarged, one with coarse grains shows blobs which limit the amount of recognizable detail. Fine grain film permits a much greater enlargement before such a limitation of detail becomes evident.

The film emulsion most widely used, so far, in holography is one having an extremely high-resolution emulsion, the Kodak 649F spectroscopic emulsion. It is capable of recording more than 1000 lines per millimeter; it could for example, record a photographically made hologram grating having 1000 lines per millimeter. Fine grain emulsions are necessary because most holograms, particularly those having off-axis reference beams, possess a very fine fringe spacing. The zone plate design in Figure 30 shows that, for it, the higher zones are at very wide off-axis angles from the focal point; their spacing approaches one wavelength of light. In holograms, similarly, if the angle between the reference beam and the widest angle of the waves from the scene exceeds sixty degrees, the fringe spacing is very close to one light wavelength. At laser wavelengths, this would be approximately 1500 lines per centimeter. It has been found that the 649F emulsion can satisfactorily accommodate this spacing; in fact, it is even satisfactory for reflection type (white light) holograms which have, as we saw in the previous chapter on reflection zone plates, exposed areas separated longitudinally within the emulsion—spaced only a half wavelength apart (3000 lines per millimeter).

The emulsion speed of 649F is only .05 (ASA rating) at the red light region of the helium-neon gas laser (6328 Angstrom units wavelength). For this emulsion and such a laser, rated at one-tenth of a watt output, an exposure time of approximately one minute is required.

During the exposure time, the interference pattern to be recorded must not be allowed to change (as would occur if relative movement took place between objects of the scene or between them and the reference beam). Accordingly, in making the hologram, the objects are usually placed on something which will not vibrate during the exposure time, such as a several-ton block of stone or marble. A shift of even 1/100,-000 of an inch of any object will change the relationship of the wave crests and troughs, thereby moving and blurring the interference pattern on the film and ruining the hologram.

If the angle between reference beam and the light from the scene is kept small, the fringe spacing becomes much larger; faster film then can be used. AGFA film Agepan FF can resolve 500 lines per millimeter and has an ASA rating of 25.

Information Content

As spectacular as is the property of holograms which provides the viewer with so much information about the scene recorded, it is also exactly that property which has unfortunately limited the use of hologram principles in many interesting applications. Such applications include three-dimensional movies and three-dimensional television, fields which would obviously benefit if the realism of holograms could be imparted to them. The large information content of a hologram is inherent in the extremely fine detail which must be recorded on the hologram film. As we saw, this detail, of about 1500 lines per millimeter, is available only in very special photographic film. Present television systems employ a far, far coarser line structure so that the outlook for using holograms in television is very bleak.

Several methods have been suggested for reducing the information content of a hologram without sacrificing completely some of its interesting properties, but this task is a very difficult one. Television pictures in the U.S. have approximately 500 lines vertically and 500 dots horizontally. One television picture (corresponding to one frame of a movie film) thus has an information content corresponding to 500 × 500, or 250,000 dots. A two-hundred millimeter by two-hundred millimeter hologram (approximately 8 inches by 8 inches) having 1500 fringes (lines) per millimeter, would have the equivalent of 200 × 1500 or 300,000 lines vertically and 300,000 dots horizontally, with a total information content corresponding to 90,000,000,000 dots (ninety *billion* dots). The ratio between the information in an 8-inch square hologram and that of a U.S. television picture is thus 360,000. To reduce the information content of a hologram by a factor that large would truly be a remarkable accomplishment.

Several procedures have been proposed to permit at least some reduction in the hologram information content, including one suggested by the author in a February 1966 publication, which involved retaining many tiny areas (perhaps 300 to 500 in each vertical line and a similar number in each horizontal line) and discarding the remaining much larger areas. Each retained area, though extremely small, would still contain several light fringes, and the assemblage would, therefore, still possess the zone plate character of the original hologram. A sequel to that proposal was investigated in 1968 by the U.S. scientist, D. J. DeBitetto, at the Philips Laboratories and L. H. Lin of the Bell Telephone Laboratories. DeBitetto retained one full, but extremely narrow, horizontal strip of a hologram (one one-hundredth of the original height), then formed a new hologram by repeating this one strip 100 times in the vertical direction. The new hologram displayed good horizontal parallax, and, accordingly, for the viewer's two, horizontally-positioned eyes, provided quite realistic three-dimensional effects. It lost, in the process, the vertical parallax originally observable by a vertical motion of the viewer.

Holograms and Coherent Radar

The last property of holography we shall discuss is one which sets holography definitely apart from photography, and shows that, in spite of the similarities, holography is not just a variant of photography, but rather, a basically new process. This fact is significant, because two relatively recent and very important new technologies, radar and sonar, have patterned themselves along the lines of the older, classical, optical science of photography, with its extensive use of lenses and reflectors. These technologies have accordingly inherited the properties and limitations of photography.

A new form of radar, coherent radar, has broken away from these traditional approaches, and it is free of the limitations of lenses and reflectors. It is free from these limitations because coherent radar is a form of holography. A wider awareness of the basic differences between holography and photography and, similarly, between coherent radar and ordinary radar, could lead to new, important developments in radar and sonar.

In a coherent radar, a generator of highly coherent microwaves provides the terrain-illuminating signal, and it also acts as a reference wave generator. As the airplane moves, the reflected signals received from each of the points along its flight path are combined with this reference signal. An interference pattern is thereby generated, and this is transformed into a light pattern and photographically recorded. This record, a form of hologram, is later processed with laser light to "reconstruct" the radar view of the terrain. Extremely fine detail is possible in the presentation of such a radar.

Coherent radar waves returning from a reflecting point have spherical wave fronts, whereas the coherent oscillator signal acts like a set of plane reference waves. Each reflecting point thus generates its own photographically recorded pattern in the form of a zone plate (a one-dimensional zone plate), similar to the two-dimensional holographic zone plate record of a point source of light. When the radar record is illuminated with laser light, an image of the original radar reflecting point (and all other reflecting points) can be reconstructed.

The comparison of holography and coherent radar underlines an important feature which is common to both and which is not possessed by ordinary optical systems; this feature is that they both provide an unlimited depth of sharp focus. In the usual optical imaging process employing a lens or parabolic reflector, only one plane section of the image field can be recorded in truly sharp focus; all other planes will be out of focus in varying degrees. Ordinary radars obey similar optical laws.

This focusing problem associated with lenses is absent both in holography and in coherent radar. This is because each point in the hologram scene (or each reflecting point in the radar field) forms its own zone plate and each recorded zone plate then causes coherent light to be sharply focused (*i.e.*, reconstructed) at exactly the proper point in space. This remarkable property of holography and coherent radar suggests that broader uses of coherent radar concepts might advantageously be made, particularly in large-aperture radars and sonars.

Extensive contributions to coherent radar were made in the late 1950s at the University of Michigan by a team under the leadership of the U.S. scientist, Louis Cutrona; one of his team members was Emmett Leith. Because the relationship between coherent radar and Gabor's holography is so close, one could almost say that holography was reinvented by Cutrona and his team. They generated zone plates (one-dimensional ones), they employed offset procedures to separate the diffracted components, and they used laser optics to regenerate the radar image. Fortunately for holography, one team member, Emmett Leith, ingeniously extended his coherent radar knowledge know-how to optical holograms, and the first visible light laser hologram thereby came into existence.

Acoustic Holograms

Akin to the microwave holograms of coherent radar are acoustic holograms, formed by recording a sound wave interference pattern. Here, too, the wave interference pattern is transformed into a light wave pattern which, after a size reduction, is viewed with laser light.

Dr. Rolf Mueller of the Bendix Research Laboratories has pioneered in using, as an ultrasonic hologram surface, the liquid-air interface above the area where the "scene" is being "illuminated" with extremely short wavelength coherent ultrasonic waves. A coherent reference wave is also directed at this liquid surface, and it becomes the "recording" surface for the hologram interference pattern. Because the surface of a liquid is a pressure-release surface, that is, a surface which "gives," or rises, at points where higher-than-average sound pressures exist, the acoustic interference pattern transforms the otherwise plane liquid surface into a surface having extremely minute, stationary ripples on it. When this rippled surface is illuminated with coherent (laser) light, an image of the submerged object is reconstructed.

Other ways of recording an acoustic interference pattern photographically, for later laser illumination, include the neon tube procedure described in connection with Figure 15; thus just as *Plate* 6 could be considered a microwave hologram, *Plate* 7 can be considered an acoustic hologram (as can the numerous other acoustic fringe patterns in this book).

PLATE 35 *(top)*. A hologram viewer using filtered white light. In the hologram reconstruction, the row of dominoes stands out behind very realistically.

PLATE 36 *(bottom)*. The Museum of Holography at 11 Mercer Street in New York City. Open to the public Wednesday through Sunday, it has many holograms on view, many of which are for sale at the Museum Bookstore.

PLATE 37. In the foreground is Dennis Gabor, the discoverer of holography. From right to left is Mrs. Gabor, Professor Jasper Holmes of the University of Hawaii, and Gilbert Devey of the National Science Foundation.

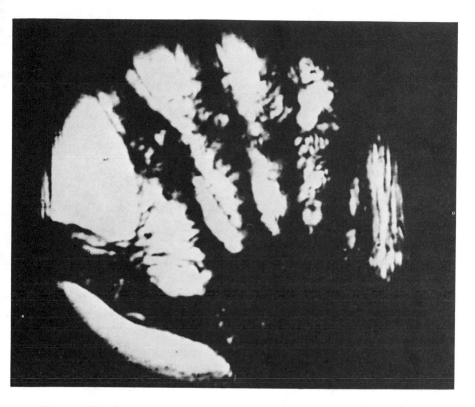

PLATE 38. A reconstructed hologram image of a human hand taken by liquid-surface acoustic holography. (Courtesy of Holosonics, Inc.)

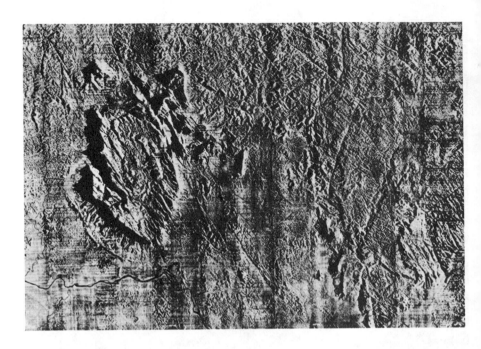

PLATE 39. A reconstructed synthetic-aperture (hologram) radar record taken during a 1971–72 survey of the Amazon River Basin. (Courtesy of Homer Jensen, Aero-Service)

PLATE 40. Placing a modern day solid-state laser on a U.S. penny demonstrates how small it is.

PLATE 41. A laser-radar picture of an airport control tower. (Courtesy of United Aircraft Research Laboratories)

Chapter 7
RECENT DEVELOPMENTS

Recently extensive improvements in and applications of lasers and holography have come about, and in this final chapter we touch briefly on some of the more important work done in the 1970s.

Viewing Holograms

Today fairly simple and inexpensive devices are available for viewing holograms even in the home (*Plate* 35), permitting the observation of not only the wonderful three-dimensional effects but also of other interesting properties, such as those present when a lens is included as one of the objects in the hologram (as in *Plate* 30). The system pictured in *Plate* 35 has been available from mail-order scientific catalogues for over five years, as have numerous holograms for viewing on that system, including real-image holograms (as discussed in connection with Figure 41). Very recently a three-dimensional movie-type hologram (moving-image, cylindrical) became available; the best-known of these is "The Kiss" by Lloyd Cross, an important contributor to holography. As the viewer moves in an arc of about 120 degrees around the illuminated cylinder, he sees the girl in the hologram first wink at him and then blow him a kiss. Another device, which has become very popular, is the holographic pendant, which uses a reflection hologram (discussed in connection with Figure 35) mounted on a 24-inch-long gold-colored chain. When illuminated with normal ambient light, it provides a brightly hued, three-dimensional image. Larger three-dimensional holograms have become popular in display advertising in airports and with such corporations as General Electric, Merck Sharpe and Dohme, and Revlon.

The Museum of Holography

December 8, 1976 was the day of the official opening of the Museum of Holography in New York (see *Plate* 36), a milestone for holography; Mayor Beame cut the ribbon for New York's eightieth museum. Among the many outstanding holography contributors present were Emmett Leith (*Plate* 27) and Anait Stephens, both of whom accepted invitations (as did the author and Dr. Stephen Benton of Polaroid) to attend the Third U.S.S.R. Conference on Holography in 1978 at Ulyanovsk on the Volga River. The museum was given a grant by the Rockefeller Foundation early in 1978 and on March 27 Dennis Gabor, the father of holography, christened the museum. Gabor was presented with museum membership card number one and accepted the position of Honorary Chairman of the Board of the Museum. Mrs. Gabor was also present at the christening. (See *Plate* 37 which the author took in Hawaii at the third U.S.-Japan seminar on holography. Gabor was a participant and the author co-chairman. Gabor died at age 78 on February 8, 1979.) In addition to its continuously changing exhibits, the museum has an excellent bookstore, which offers a wide selection of holograms.

White Light or "Rainbow" Holograms

Recently a new type of hologram was developed,; Benton is generally regarded as its father. It is made in two steps. First a hologram of a scene is recorded by using a laser. It is then masked so that only a thin horizontal stripe remains transparent. (This is comparable to blocking out all but a thin horizontal stripe of the zone plate hologram of *Plate* 5.) The masked hologram is now illuminated to reconstruct the virtual and real images, as was shown in Figure 9, but only the (focussed) real image (shown as "discarded" in Figure 9) is used to make the second, final hologram. If normal illumination of the first, masked, hologram were used, the real image as generated would be pseudoscopic, as noted in connection with Figure 40. Therefore, the *phase-conjugate* of the original reference beam is used for illumination, resulting in a pseudoscopic-pseudoscopic image (a doubly-reversed image; see Chapter 6). Also, because of the narrow vertical dimension of the masked hologram, all vertical perspective information is lost. (Similarly, the vertical focussing action of the zone plate of *Plate 5* is eliminated when only a horizontal stripe of it is available.) When this second hologram (made from the real image of the masked first one) is viewed with laser light, a real image of the scene is reconstructed, but only at the vertical position occupied by the slit formed by the original mask. The full hologram effect is created, including the capacity to look around objects in the foreground, by moving from side to side (as in *Plate* 26). But because only the *slit* area is reconstructed, the light and the image disappear when the viewer's eyes move above or below the slit image.

At this point a white light reconstructing source is used in place of the laser. As noted in connection with Figure 29, the diffraction angle is different for waves of different wavelengths (light of different colors). For rainbow holograms as well the vertical direction in which the slit and the image are reconstructed depends on the color of the reconstructing light. With white light, all colors are present; thus as the viewer moves up and down, the image is seen in different colors. Each level provides, however, the full three-dimensional view. Since demonstrating this first form of rainbow hologram, Benton has added new steps to produce excellent achromatic images (practically devoid of color). His best-known is a holographic reproduction of the Boston Museum of Fine Arts sculpture of Aphrodite, which the Museum of Holography's newsletter *Holosphere* (December 1978) says is "as white as the marble original."

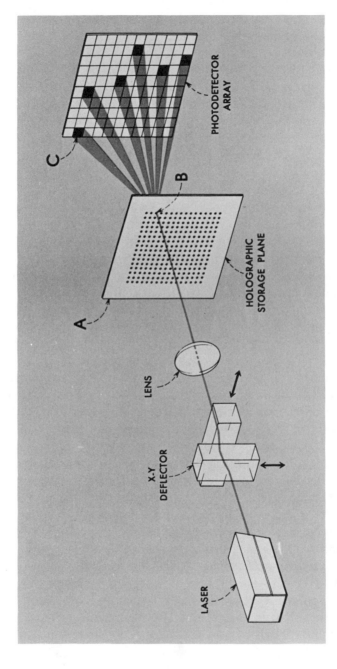

Figure 43. Sketch of a flying-spot holographic computer memory. (Courtesy of Bell Laboratories)

Hologram Computer Memories

It was recognized early in the history of holography that a hologram has a very large information storage capability, potentially offering significant improvements in capacity and access time over existing computer mass-memory technologies. The main interest in optical memories stems from the large bit densities that are attainable. Lasers offer a large increase in bit packing density over magnetic recording since the bit size can approach the wavelength of light (0.5 μm). Today hologram memories are primarily directed at mass storage applications.

Figure 43 shows an example of a technique called a holographic flying-spot scanning memory. The X-Y deflector directs a laster beam to one of the stored holograms, and the lasser light emerging from a hologram illuminates certain ones of the array of photodetectors in accordance with the information that had previously been stored in the holographic memory. Memories of the type shown in Figure 43 are referred to as *read-only memories* (ROM) and they are used where large amounts of permanent data must be stored, or where only infrequent changes are needed. The optical storage of 400 pages of information in one square inch has been demonstrated, and trillion bit (10^{12}) laser memories have been applied at the American Oil Company and at the University of Illinois for use with the Advanced Research Projects Agency's Illiac IV computer.

Liquid Surface Holography

Liquid surface acoustic holography technique has developed very rapidly and commercial equipment is now available. Figure 44 shows the principles involved. The advantages in using a liquid surface instead of photographic film are that no development is required and the liquid surface responds rapidly to the ultrasonic energy, thereby permitting an instantaneous readout. In Figure 44 the inter-

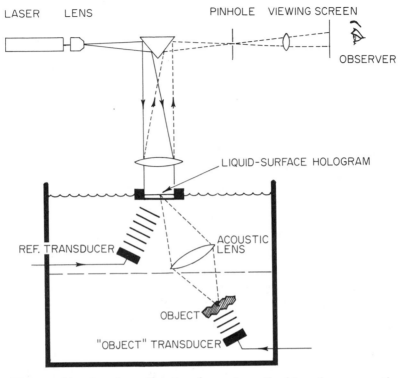

LASER   LENS   PINHOLE   VIEWING SCREEN

OBSERVER

LIQUID-SURFACE HOLOGRAM

REF. TRANSDUCER

ACOUSTIC LENS

OBJECT

"OBJECT" TRANSDUCER

Figure 44. A schematic drawing of a liquid-surface acoustic holography system.

action of sound with the liquid surface is such that, where the object has good transmission, a strong ripple pattern will develop on the liquid surface because of the interference of the object information with the reference beam wave. Where the ripple pattern is strong, much light will be diffracted from the main beam; where the transmission through the object is poor, the ripple pattern will be weak.

Plate 38 shows an example of the reconstruction of a liquid surface hologram when a hand is the object. Such images, produced through transmission, are primarily the result of the absorption characteristics of the object. The ability to detect small breast tumors may prove to be the greatest service of liquid surface holography.

Synthetic-Aperture Radar and Sonar

One of the most extensive uses of nonoptical holograms has occurred in the microwave radar field in the form of synthetic-aperture radar. Shortly before the announcement of his award of the 1971 Nobel Prize in Physics, Gabor commented, "Unknown to me, a most interesting branch of holography was developing from 1956 onwards at the Willow Run Laboratory attached to the University of Michigan. It was holography with electromagnetic waves, and reconstruction by light, which was called 'Side Looking Radar' or 'Synthetic Aerials.' It was classified work; the first publication by Cutrona, Leith, Palermo and Porcello occurred in 1960. Reconstructions of the object plane by illumination with a monochromatic mercury lamp were of impressive perfection."

Leith was one of the principal contributors to synthetic-aperture development and, with Upatnieks, was the first to use the laser in holography. Gabor commented, "I came to [holography] through the electron microscope, [Leith] through side-looking coherent radar." In coherent radar systems, an aircraft moving along a very straight path continually emits successive microwave pulses. The frequency of the microwave signal is very constant (that is, the signal remains coherent with itself for very long periods). During these periods the aircraft may have traveled several thousand feet, but because the signals are coherent, all of the many echoes that return during this period can be processed as though a single antenna as long as the flight path had been used. The effective antenna size is thus quite large. As in the optical case, where the larger the optical telescope the greater the detail possible, this large "synthetic" aperture provides reconstructed (hologram-type) records with extremely fine detail. *Plate* 39 shows such a record of the Amazon River Basin, an area difficult to photograph because of the almost constant clouds, but through which the radar's microwaves pass easily.

Because of its close relationship to radar, sonar ("acoustic radar") has also benefitted from holography. An example of this is found in B-scan sonar, a beneficial technique used in medical examinations.

New Lasers

Recent developments in lasers include huge, successful ones, such as the Argus laser at the University of California Lawrence Livermore Laboratory (which can produce heat reaching 100 million degrees), and extremely tiny ones, such as the heterojunction solid-state laser shown in *Plate* 40. Hopes are high that the huge lasers, such as the Argus and the newer, bigger, $25 million Shiva system, will permit controlled nuclear fusion, the clean energy hope for the future. For fuel, fusion will draw on the great quantity of heavy hydrogen in sea water. The tiny, low-power but highly efficient, solid-state lasers have already found a place in the area of communication with light waves, a field of rapid growth. Medium-power lasers also have a rapidly expanding field, that of marking products quickly and permanently. Already these lasers are being used extensively to mark bearings, wire, auto parts, and electronic components, among other products.

Wires of Glass

Until recently, communication over long distances was by wire, coaxial cable, microwave radio relay, or satellite. The last two of these carry the widest bandwidths (the most channels), but they must use microwave frequencies in order to avoid losses from absorption caused by clouds and rain. Recently, very low-loss optical fibers (wire or glass) have been developed, and with lasers' high light-wave frequencies, extremely wide communications bandwidths (many, many channels) are now possible. The British Post Office has announced that it was awarding contracts for 450 kilometers of such optical links at a cost of $13 million, with a data rate (a measure of the bandwidths possible) of 140 megabits per second. The American Telephone and Telegraph Company has installed several fiber optics communications links, including several in New York State for video links for national television coverage of the 1980 Olympic Games in Lake Placid. Many companies now offer commercial low-loss optical-fiber connectors. It has been predicted that fiber optics will become a $1 billion industry in the 1980s.

Lasers in the Military

The military uses lasers for range-finding, in laser-guided bombs, for guiding projectiles by the infantry, and for the Marines' lightweight laser "guns." One of the more interesting developments is light-wave radar, often called *lidar, Plate* 41 shows how clearly such a "radar" can delineate and identify structures. *Star Wars* may not be pure fantasy!

EPILOGUE

Probably the most intriguing aspect of holography is the way in which all of the parts of the quite complicated puzzle came together so elegantly for Gabor. It suggests an inventiveness and a clairvoyance not often encountered.

Gabor's first step was to recognize that all properties of a wave pattern formed at a plane (*e.g.,* the plane of a window) by single wavelength light issuing from a scene are completely specified once the size and relative positions of wave crests and troughs (the amplitudes and relative *phases*) are specified. (Even this initial basic hypothesis of Gabor's was, as we have seen, questioned by some.) This recognition was, however, just the first step; to exploit it, he had to find, not only a way of *recording* both amplitude and crest position (phase) of the pattern, but also a way of *retrieving,* from such a record, the original wave pattern.

Let us first think about the recording procedure. One very common recording technique employs a device having a pen which swings back and forth on a moving strip of paper, thereby recording the electrical waves which cause the pen to move. Such a device *could* have recorded the variations in the output of the microwave amplifier during the scanning process of Figure 15 (in place of the neon tube and camera technique employed). Had this been done, a paper record of the complete microwave interference pattern (containing both amplitude and phase) would have been obtained. Although all the pertinent information would thus have been recorded, the *useful retrieval* of the original wave pattern would have posed a difficult problem.

Gabor recorded his wave patterns photographically. His

hologram records also contained amplitude and phase information about the wave pattern, but, as in the case of the paper record, for the information to be retrieved so as to generate a replica of the wave pattern, a third step was required—one far from obvious. In recent years, the problem of usefully retrieving information which has been recorded (as, for example, in thousands of books in a library) has been attacked with great vigor. If someone had suggested to an information-retrieval expert that 36 billion dots of information in a photographic recording could be reassembled so as to re-create a prior light wave pattern simply by shining coherent light on the record, he would surely have been most skeptical.

Gabor wrote down two equations, which Stroke has called "the two basic Gabor equations." From his first equation, he knew that the wave amplitude and relative crest positions would truly be recorded when the scene waves were combined with a reference wave. We saw in Figure 7 that phase information *is* recorded because the light is bright at the $(++)$ points and less bright at the $(+-)$ points.

At this point, we noted that Gabor's recording process possessed one inherent flaw: the hologram record of phase and amplitude does *not* have a one-to-one correspondence with the *desired* recorded wave set. Thus, in the photographic record of the point source and plane reference wave of Figure 7, an identical phase record would have been produced by several *other* wave patterns. We can see this by imagining that the blocking zones of the zone plate of Figure 30 are reflecting rather than absorbing. Plane waves arriving from the left would then generate, in addition to the three wave sets (converging, diverging, and straight-through) on the right, a similar group of three *reflected* wave sets on the left. If a wave set similar to any one of these four reconstructed converging and diverging wave sets were to have been used in the original recording process (in conjunction with a plane reference wave set), the identical (planar) hologram zone plate would have resulted. The situation is comparable to that of recording a word which has numerous meanings, and then trying to avoid ambiguity in the replay and ascertaining *which* of the meanings was intended in the recording.

This ambiguity could easily have daunted another investigator, but it did not daunt Gabor. When he set down his second equation, expressing mathematically the interaction between the recorded pattern (the hologram) and a third wave set (actually, one identical to the reference wave), he realized that wave sets *other* than the desired set would also be reconstructed. Nevertheless, the fact that the desired set *would indeed* be reconstructed was of far greater significance.

Gabor subsequently proved experimentally the validity of his conclusions, but in order to achieve maximum image brightness with his weak source of coherent light, he viewed his images by looking directly toward the source; the resulting superposition of images caused some blurring. Sixteen years later, Leith was able, with the far more powerful light from a laser, to exploit the offset procedure of coherent radar, thereby angularly separating the several components. The puzzle was assembled, the mission accomplished, and our story ends.

SUGGESTED READING

R. J. Coliler, C. B. Burckhardt, and L. H. Lin, OPTICAL HOLOGRAPHY. Academic Press, 1971.

M. Françon, HOLOGRAPHY. Academic Press, 1971.

P. Greguss, ed., HOLOGRAPHY IN MEDICINE. I.P.C. Science and Technology Press, 1975.

B. P. Hildebrand and B. B. Brendon, AN INTRODUCTION TO ACOUSTICAL HOLOGRAPHY. Plenum Press, 1972.

Winston E. Kock, THE CREATIVE ENGINEER. Plenum Press, 1978.

Winston E. Kock, ENGINEERING APPLICATIONS OF LASERS AND HOLOGRAPHY. Plenum Press, 1975.

Winston E. Kock, RADAR, SONAR, AND HOLOGRAPHY. Academic Press, 1973.

Winston E. Kock, SOUND WAVES AND LIGHT WAVES: THE FUNDAMENTALS OF WAVE MOTION, Anchor Science Study Series, Doubleday & Company, 1965.

Bela A. Lengyel, LASERS. 2d ed., Wiley Interscience, 1971.

Edward Renchardt, LIGHT, VISIBLE AND INVISIBLE. University of Michigan Press, 1965.

George W. Stoke, AN INTRODUCTION TO COHERENT OPTICS AND HOLOGRAPHY. 2d ed., Academic Press, 1969.

INDEX

WINSTON E. KOCK received his Electrical Engineering degree from the University of Cincinnati in 1932 and his M.S. degree in Physics the following year. From Cincinnati, he went to the University of Berlin, in Germany, where he received his Ph.D. in Physics *cum laude* in 1934. In Berlin, he came in contact with Max Planck and Max von Laue, two of the great men in science in the early years of this century. After a year as a teaching fellow at Cincinnati, he attended the Institute for Advanced Study at Princeton, New Jersey, and there took courses under Albert Einstein, John von Neumann, and Eugene Wigner, among others. In 1936, he spent a summer at the Indian Institute of Science at Bangalore, India, studying under Sir C. V. Raman, a Nobel Laureate. It was during this summer that his interest in optics was first stimulated—an interest that was to grow through the years, leading to the research on which Dr. Kock has based LASERS AND HOLOGRAPHY.

An unusually varied career in engineering and science has led to numerous contributions to acoustics, radar, solid state physics, and television, and Dr. Kock devoted several years to the development of a commercial version of the electronic organ (which he originally designed in 1932 for his senior thesis at the University of Cincinnati) for the Baldwin Piano Company. In 1942, he joined the Radio Research Department of Bell Telephone Laboratories at Holmdel, New Jersey, where he worked, under Harald T. Friis, on microwave lenses which eventually were to go into the Bell System's coast-to-coast microwave relay circuits. He transferred to the Bell Murray Hill Laboratories in 1948 to work in acoustics and on the transistor (he holds a joint patent on the coaxial transistor). He became Director of Acoustics Research at Bell Laboratories, invented several lenses for loudspeakers, and worked on the computer for speech recognition. He also directed research on a narrow-band television transmission system based on acoustic techniques. In 1956, he joined the Bendix Corporation, where he was Vice President and Chief

Scientist until his retirement in 1971; now he serves as a consultant. From 1964 to 1966, Dr. Kock was the first Director of the NASA Electronics Research Center. He is currently Visiting Professor and Director of the Herman Schneider Research Laboratory at the University of Cincinnati.

Dr. Kock has received many honors, among them the Navy's highest civilian award, the Distinguished Public Service Medal (1964); an Honorary Fellowship in the Indian Academy of Sciences (1970); an honorary D.Sc. from the University of Cincinnati (1952); Eta Kappa Nu's Outstanding Young Electrical Engineer Award (1938) and its Eminent Member Award (1966); fellowships in the Acoustical Society, the Physical Society, and the L.E.E.E.; and membership in Tau Beta Pi, Sigma Xi, and Eta Kappa Nu. He was on the Governing Board of the American Institute of Physics (1957–63) and was chairman of the Professional Group on Audio of the Institute of Radio Engineers (1954–55). He is a member of the Board of the Roanwell Corporation and Hadron, Inc., and has been Chairman of the Board of Trustees of Western College for Women and Board member of Argonne Universities Association and the Atomic Industrial Forum.

Holder of over eighty patents, Dr. Kock is also the author of SOUND WAVES AND LIGHT WAVES (1965), SEEING SOUND (1971), RADAR, SONAR, AND HOLOGRAPHY (1973), ENGINEERING APPLICATIONS OF LASERS AND HOLOGRAPHY (1973), and THE CREATIVE ENGINEER (1978).

A CATALOG OF SELECTED
DOVER BOOKS
IN ALL FIELDS OF INTEREST

A CATALOG OF SELECTED

DOVER BOOKS

IN ALL FIELDS OF INTEREST

DRAWINGS OF REMBRANDT, edited by Seymour Slive. Updated Lippmann, Hofstede de Groot edition, with definitive scholarly apparatus. All portraits, biblical sketches, landscapes, nudes. Oriental figures, classical studies, together with selection of work by followers. 550 illustrations. Total of 630pp. 9⅛ × 12¼.
21485-0, 21486-9 Pa., Two-vol. set $29.90

GHOST AND HORROR STORIES OF AMBROSE BIERCE, Ambrose Bierce. 24 tales vividly imagined, strangely prophetic, and decades ahead of their time in technical skill: "The Damned Thing," "An Inhabitant of Carcosa," "The Eyes of the Panther," "Moxon's Master," and 20 more. 199pp. 5⅜ × 8½. 20767-6 Pa. $4.95

ETHICAL WRITINGS OF MAIMONIDES, Maimonides. Most significant ethical works of great medieval sage, newly translated for utmost precision, readability. Laws Concerning Character Traits, Eight Chapters, more. 192pp. 5⅜ × 8½.
24522-5 Pa. $5.95

THE EXPLORATION OF THE COLORADO RIVER AND ITS CANYONS, J. W. Powell. Full text of Powell's 1,000-mile expedition down the fabled Colorado in 1869. Superb account of terrain, geology, vegetation, Indians, famine, mutiny, treacherous rapids, mighty canyons, during exploration of last unknown part of continental U.S. 400pp. 5⅜ × 8½. 20094-9 Pa. $7.95

HISTORY OF PHILOSOPHY, Julián Marías. Clearest one-volume history on the market. Every major philosopher and dozens of others, to Existentialism and later. 505pp. 5⅜ × 8½. 21739-6 Pa. $9.95

ALL ABOUT LIGHTNING, Martin A. Uman. Highly readable nontechnical survey of nature and causes of lightning, thunderstorms, ball lightning, St. Elmo's Fire, much more. Illustrated. 192pp. 5⅜ × 8½. 25237-X Pa. $5.95

SAILING ALONE AROUND THE WORLD, Captain Joshua Slocum. First man to sail around the world, alone, in small boat. One of great feats of seamanship told in delightful manner. 67 illustrations. 294pp. 5⅜ × 8½. 20326-3 Pa. $4.95

LETTERS AND NOTES ON THE MANNERS, CUSTOMS AND CONDITIONS OF THE NORTH AMERICAN INDIANS, George Catlin. Classic account of life among Plains Indians: ceremonies, hunt, warfare, etc. 312 plates. 572pp. of text. 6⅛ × 9¼. 22118-0, 22119-9, Pa., Two-vol. set $17.90

THE SECRET LIFE OF SALVADOR DALÍ, Salvador Dalí. Outrageous but fascinating autobiography through Dalí's thirties with scores of drawings and sketches and 80 photographs. A must for lovers of 20th-century art. 432pp. 6½ × 9¼. (Available in U.S. only) 27454-3 Pa. $9.95

THE BOOK OF BEASTS: Being a Translation from a Latin Bestiary of the Twelfth Century, T. H. White. Wonderful catalog of real and fanciful beasts: manticore, griffin, phoenix, amphivius, jaculus, many more. White's witty erudite commentary on scientific, historical aspects enhances fascinating glimpse of medieval mind. Illustrated. 296pp. 5⅜ × 8¼. (Available in U.S. only)   24609-4 Pa. $7.95

FRANK LLOYD WRIGHT: Architecture and Nature with 160 Illustrations, Donald Hoffmann. Profusely illustrated study of influence of nature—especially prairie—on Wright's designs for Fallingwater, Robie House, Guggenheim Museum, other masterpieces. 96pp. 9¼ × 10¾.   25098-9 Pa. $8.95

FRANK LLOYD WRIGHT'S FALLINGWATER, Donald Hoffmann. Wright's famous waterfall house: planning and construction of organic idea. History of site, owners, Wright's personal involvement. Photographs of various stages of building. Preface by Edgar Kaufmann, Jr. 100 illustrations. 112pp. 9¼ × 10.
                                                                          23671-4 Pa. $8.95

YEARS WITH FRANK LLOYD WRIGHT: Apprentice to Genius, Edgar Tafel. Insightful memoir by a former apprentice presents a revealing portrait of Wright the man, the inspired teacher, the greatest American architect. 372 black-and-white illustrations. Preface. Index. vi + 228pp. 8¼ × 11.   24801-1 Pa. $10.95

THE STORY OF KING ARTHUR AND HIS KNIGHTS, Howard Pyle. Enchanting version of King Arthur fable has delighted generations with imaginative narratives of exciting adventures and unforgettable illustrations by the author. 41 illustrations. xviii + 313pp. 6⅛ × 9¼.   21445-1 Pa. $6.95

THE GODS OF THE EGYPTIANS, E. A. Wallis Budge. Thorough coverage of numerous gods of ancient Egypt by foremost Egyptologist. Information on evolution of cults, rites and gods; the cult of Osiris; the Book of the Dead and its rites; the sacred animals and birds; Heaven and Hell; and more. 956pp. 6⅛ × 9¼.
                                        22055-9, 22056-7 Pa., Two-vol. set $21.90

A THEOLOGICO-POLITICAL TREATISE, Benedict Spinoza. Also contains unfinished *Political Treatise*. Great classic on religious liberty, theory of government on common consent. R. Elwes translation. Total of 421pp. 5⅜ × 8½.
                                                                          20249-6 Pa. $7.95

INCIDENTS OF TRAVEL IN CENTRAL AMERICA, CHIAPAS, AND YUCATAN, John L. Stephens. Almost single-handed discovery of Maya culture; exploration of ruined cities, monuments, temples; customs of Indians. 115 drawings. 892pp. 5⅜ × 8½.   22404-X, 22405-8 Pa., Two-vol. set $17.90

LOS CAPRICHOS, Francisco Goya. 80 plates of wild, grotesque monsters and caricatures. Prado manuscript included. 183pp. 6⅜ × 9⅜.   22384-1 Pa. $6.95

AUTOBIOGRAPHY: The Story of My Experiments with Truth, Mohandas K. Gandhi. Not hagiography, but Gandhi in his own words. Boyhood, legal studies, purification, the growth of the Satyagraha (nonviolent protest) movement. Critical, inspiring work of the man who freed India. 480pp. 5⅜ × 8½. (Available in U.S. only)
                                                                          24593-4 Pa. $6.95

ILLUSTRATED DICTIONARY OF HISTORIC ARCHITECTURE, edited by Cyril M. Harris. Extraordinary compendium of clear, concise definitions for over 5,000 important architectural terms complemented by over 2,000 line drawings. Covers full spectrum of architecture from ancient ruins to 20th-century Modernism. Preface. 592pp. 7½ × 9⅜. 24444-X Pa. $15.95

THE NIGHT BEFORE CHRISTMAS, Clement Moore. Full text, and woodcuts from original 1848 book. Also critical, historical material. 19 illustrations. 40pp. 4⅝ × 6. 22797-9 Pa. $2.50

THE LESSON OF JAPANESE ARCHITECTURE: 165 Photographs, Jiro Harada. Memorable gallery of 165 photographs taken in the 1930's of exquisite Japanese homes of the well-to-do and historic buildings. 13 line diagrams. 192pp. 8⅝ × 11¼. 24778-3 Pa. $10.95

THE AUTOBIOGRAPHY OF CHARLES DARWIN AND SELECTED LETTERS, edited by Francis Darwin. The fascinating life of eccentric genius composed of an intimate memoir by Darwin (intended for his children); commentary by his son, Francis; hundreds of fragments from notebooks, journals, papers; and letters to and from Lyell, Hooker, Huxley, Wallace and Henslow. xi + 365pp. 5⅜ × 8. 20479-0 Pa. $6.95

WONDERS OF THE SKY: Observing Rainbows, Comets, Eclipses, the Stars and Other Phenomena, Fred Schaaf. Charming, easy-to-read poetic guide to all manner of celestial events visible to the naked eye. Mock suns, glories, Belt of Venus, more. Illustrated. 299pp. 5¼ × 8¼. 24402-4 Pa. $7.95

BURNHAM'S CELESTIAL HANDBOOK, Robert Burnham, Jr. Thorough guide to the stars beyond our solar system. Exhaustive treatment. Alphabetical by constellation: Andromeda to Cetus in Vol. 1; Chamaeleon to Orion in Vol. 2; and Pavo to Vulpecula in Vol. 3. Hundreds of illustrations. Index in Vol. 3. 2,000pp. 6⅛ × 9¼. 23567-X, 23568-8, 23673-0 Pa., Three-vol. set $41.85

STAR NAMES: Their Lore and Meaning, Richard Hinckley Allen. Fascinating history of names various cultures have given to constellations and literary and folkloristic uses that have been made of stars. Indexes to subjects. Arabic and Greek names. Biblical references. Bibliography. 563pp. 5⅜ × 8½. 21079-0 Pa. $8.95

THIRTY YEARS THAT SHOOK PHYSICS: The Story of Quantum Theory, George Gamow. Lucid, accessible introduction to influential theory of energy and matter. Careful explanations of Dirac's anti-particles, Bohr's model of the atom, much more. 12 plates. Numerous drawings. 240pp. 5⅜ × 8½. 24895-X Pa. $5.95

CHINESE DOMESTIC FURNITURE IN PHOTOGRAPHS AND MEASURED DRAWINGS, Gustav Ecke. A rare volume, now affordably priced for antique collectors, furniture buffs and art historians. Detailed review of styles ranging from early Shang to late Ming. Unabridged republication. 161 black-and-white drawings, photos. Total of 224pp. 8⅜ × 11¼. (Available in U.S. only) 25171-3 Pa. $13.95

VINCENT VAN GOGH: A Biography, Julius Meier-Graefe. Dynamic, penetrating study of artist's life, relationship with brother, Theo, painting techniques, travels, more. Readable, engrossing. 160pp. 5⅜ × 8½. (Available in U.S. only) 25253-1 Pa. $4.95

CATALOG OF DOVER BOOKS

HOW TO WRITE, Gertrude Stein. Gertrude Stein claimed anyone could understand her unconventional writing—here are clues to help. Fascinating improvisations, language experiments, explanations illuminate Stein's craft and the art of writing. Total of 414pp. 4⅝ × 6⅜. 23144-5 Pa. $6.95

ADVENTURES AT SEA IN THE GREAT AGE OF SAIL: Five Firsthand Narratives, edited by Elliot Snow. Rare true accounts of exploration, whaling, shipwreck, fierce natives, trade, shipboard life, more. 33 illustrations. Introduction. 353pp. 5⅜ × 8½. 25177-2 Pa. $8.95

THE HERBAL OR GENERAL HISTORY OF PLANTS, John Gerard. Classic descriptions of about 2,850 plants—with over 2,700 illustrations—includes Latin and English names, physical descriptions, varieties, time and place of growth, more. 2,706 illustrations. xlv + 1,678pp. 8½ × 12¼. 23147-X Cloth. $75.00

DOROTHY AND THE WIZARD IN OZ, L. Frank Baum. Dorothy and the Wizard visit the center of the Earth, where people are vegetables, glass houses grow and Oz characters reappear. Classic sequel to Wizard of Oz. 256pp. 5⅜ × 8. 24714-7 Pa. $5.95

SONGS OF EXPERIENCE: Facsimile Reproduction with 26 Plates in Full Color, William Blake. This facsimile of Blake's original "Illuminated Book" reproduces 26 full-color plates from a rare 1826 edition. Includes "The Tyger," "London," "Holy Thursday," and other immortal poems. 26 color plates. Printed text of poems. 48pp. 5¼ × 7. 24636-1 Pa. $3.95

SONGS OF INNOCENCE, William Blake. The first and most popular of Blake's famous "Illuminated Books," in a facsimile edition reproducing all 31 brightly colored plates. Additional printed text of each poem. 64pp. 5¼ × 7. 22764-2 Pa. $3.95

PRECIOUS STONES, Max Bauer. Classic, thorough study of diamonds, rubies, emeralds, garnets, etc.: physical character, occurrence, properties, use, similar topics. 20 plates, 8 in color. 94 figures. 659pp. 6⅛ × 9¼. 21910-0, 21911-9 Pa., Two-vol. set $15.90

ENCYCLOPEDIA OF VICTORIAN NEEDLEWORK, S. F. A. Caulfeild and Blanche Saward. Full, precise descriptions of stitches, techniques for dozens of needlecrafts—most exhaustive reference of its kind. Over 800 figures. Total of 679pp. 8⅜ × 11. Two volumes. Vol. 1 22800-2 Pa. $11.95 Vol. 2 22801-0 Pa. $11.95

THE MARVELOUS LAND OF OZ, L. Frank Baum. Second Oz book, the Scarecrow and Tin Woodman are back with hero named Tip, Oz magic. 136 illustrations. 287pp. 5⅜ × 8½. 20692-0 Pa. $5.95

WILD FOWL DECOYS, Joel Barber. Basic book on the subject, by foremost authority and collector. Reveals history of decoy making and rigging, place in American culture, different kinds of decoys, how to make them, and how to use them. 140 plates. 156pp. 7⅞ × 10¾. 20011-6 Pa. $8.95

HISTORY OF LACE, Mrs. Bury Palliser. Definitive, profusely illustrated chronicle of lace from earliest times to late 19th century. Laces of Italy, Greece, England, France, Belgium, etc. Landmark of needlework scholarship. 266 illustrations. 672pp. 6⅛ × 9¼. 24742-2 Pa. $14.95

ILLUSTRATED GUIDE TO SHAKER FURNITURE, Robert Meader. All furniture and appurtenances, with much on unknown local styles. 235 photos. 146pp. 9 × 12. 22819-3 Pa. $8.95

WHALE SHIPS AND WHALING: A Pictorial Survey, George Francis Dow. Over 200 vintage engravings, drawings, photographs of barks, brigs, cutters, other vessels. Also harpoons, lances, whaling guns, many other artifacts. Comprehensive text by foremost authority. 207 black-and-white illustrations. 288pp. 6 × 9. 24808-9 Pa. $9.95

THE BERTRAMS, Anthony Trollope. Powerful portrayal of blind self-will and thwarted ambition includes one of Trollope's most heartrending love stories. 497pp. 5⅜ × 8½. 25119-5 Pa. $9.95

ADVENTURES WITH A HAND LENS, Richard Headstrom. Clearly written guide to observing and studying flowers and grasses, fish scales, moth and insect wings, egg cases, buds, feathers, seeds, leaf scars, moss, molds, ferns, common crystals, etc.—all with an ordinary, inexpensive magnifying glass. 209 exact line drawings aid in your discoveries. 220pp. 5⅜ × 8½. 23330-8 Pa. $4.95

RODIN ON ART AND ARTISTS, Auguste Rodin. Great sculptor's candid, wide-ranging comments on meaning of art; great artists; relation of sculpture to poetry, painting, music; philosophy of life, more. 76 superb black-and-white illustrations of Rodin's sculpture, drawings and prints. 119pp. 8⅜ × 11¼. 24487-3 Pa. $7.95

FIFTY CLASSIC FRENCH FILMS, 1912–1982: A Pictorial Record, Anthony Slide. Memorable stills from Grand Illusion, Beauty and the Beast, Hiroshima, Mon Amour, many more. Credits, plot synopses, reviews, etc. 160pp. 8¼ × 11. 25256-6 Pa. $11.95

THE PRINCIPLES OF PSYCHOLOGY, William James. Famous long course complete, unabridged. Stream of thought, time perception, memory, experimental methods; great work decades ahead of its time. 94 figures. 1,391pp. 5⅜ × 8½. 20381-6, 20382-4 Pa., Two-vol. set $23.90

BODIES IN A BOOKSHOP, R. T. Campbell. Challenging mystery of blackmail and murder with ingenious plot and superbly drawn characters. In the best tradition of British suspense fiction. 192pp. 5⅜ × 8½. 24720-1 Pa. $4.95

CALLAS: PORTRAIT OF A PRIMA DONNA, George Jellinek. Renowned commentator on the musical scene chronicles incredible career and life of the most controversial, fascinating, influential operatic personality of our time. 64 black-and-white photographs. 416pp. 5⅜ × 8¼. 25047-4 Pa. $8.95

GEOMETRY, RELATIVITY AND THE FOURTH DIMENSION, Rudolph Rucker. Exposition of fourth dimension, concepts of relativity as Flatland characters continue adventures. Popular, easily followed yet accurate, profound. 141 illustrations. 133pp. 5⅜ × 8½. 23400-2 Pa. $4.95

HOUSEHOLD STORIES BY THE BROTHERS GRIMM, with pictures by Walter Crane. 53 classic stories—Rumpelstiltskin, Rapunzel, Hansel and Gretel, the Fisherman and his Wife, Snow White, Tom Thumb, Sleeping Beauty, Cinderella, and so much more—lavishly illustrated with original 19th century drawings. 114 illustrations. x + 269pp. 5⅜ × 8½. 21080-4 Pa. $4.95

CATALOG OF DOVER BOOKS

SUNDIALS, Albert Waugh. Far and away the best, most thorough coverage of ideas, mathematics concerned, types, construction, adjusting anywhere. Over 100 illustrations. 230pp. 5⅜ × 8½. 22947-5 Pa. $5.95

PICTURE HISTORY OF THE NORMANDIE: With 190 Illustrations, Frank O. Braynard. Full story of legendary French ocean liner: Art Deco interiors, design innovations, furnishings, celebrities, maiden voyage, tragic fire, much more. Extensive text. 144pp. 8⅞ × 11¾. 25257-4 Pa. $10.95

THE FIRST AMERICAN COOKBOOK: A Facsimile of "American Cookery," 1796, Amelia Simmons. Facsimile of the first American-written cookbook published in the United States contains authentic recipes for colonial favorites—pumpkin pudding, winter squash pudding, spruce beer, Indian slapjacks, and more. Introductory Essay and Glossary of colonial cooking terms. 80pp. 5⅜ × 8½. 24710-4 Pa. $3.50

101 PUZZLES IN THOUGHT AND LOGIC, C. R. Wylie, Jr. Solve murders and robberies, find out which fishermen are liars, how a blind man could possibly identify a color—purely by your own reasoning! 107pp. 5⅜ × 8½. 20367-0 Pa. $2.95

ANCIENT EGYPTIAN MYTHS AND LEGENDS, Lewis Spence. Examines animism, totemism, fetishism, creation myths, deities, alchemy, art and magic, other topics. Over 50 illustrations. 432pp. 5⅜ × 8½. 26525-0 Pa. $8.95

ANTHROPOLOGY AND MODERN LIFE, Franz Boas. Great anthropologist's classic treatise on race and culture. Introduction by Ruth Bunzel. Only inexpensive paperback edition. 255pp. 5⅜ × 8½. 25245-0 Pa. $7.95

THE TALE OF PETER RABBIT, Beatrix Potter. The inimitable Peter's terrifying adventure in Mr. McGregor's garden, with all 27 wonderful, full-color Potter illustrations. 55pp. 4¼ × 5½. (Available in U.S. only) 22827-4 Pa. $1.75

THREE PROPHETIC SCIENCE FICTION NOVELS, H. G. Wells. *When the Sleeper Wakes, A Story of the Days to Come* and *The Time Machine* (full version). 335pp. 5⅜ × 8½. (Available in U.S. only) 20605-X Pa. $8.95

APICIUS COOKERY AND DINING IN IMPERIAL ROME, edited and translated by Joseph Dommers Vehling. Oldest known cookbook in existence offers readers a clear picture of what foods Romans ate, how they prepared them, etc. 49 illustrations. 301pp. 6⅛ × 9¼. 23563-7 Pa. $7.95

SHAKESPEARE LEXICON AND QUOTATION DICTIONARY, Alexander Schmidt. Full definitions, locations, shades of meaning of every word in plays and poems. More than 50,000 exact quotations. 1,485pp. 6½ × 9¼. 22726-X, 22727-8 Pa., Two-vol. set $31.90

THE WORLD'S GREAT SPEECHES, edited by Lewis Copeland and Lawrence W. Lamm. Vast collection of 278 speeches from Greeks to 1970. Powerful and effective models; unique look at history. 842pp. 5⅜ × 8½. 20468-5 Pa. $12.95

THE BLUE FAIRY BOOK, Andrew Lang. The first, most famous collection, with many familiar tales: Little Red Riding Hood, Aladdin and the Wonderful Lamp, Puss in Boots, Sleeping Beauty, Hansel and Gretel, Rumpelstiltskin; 37 in all. 138 illustrations. 390pp. 5⅜ × 8½. 21437-0 Pa. $6.95

THE STORY OF THE CHAMPIONS OF THE ROUND TABLE, Howard Pyle. Sir Launcelot, Sir Tristram and Sir Percival in spirited adventures of love and triumph retold in Pyle's inimitable style. 50 drawings, 31 full-page. xviii + 329pp. 6½ × 9¼. 21883-X Pa. $7.95

THE MYTHS OF THE NORTH AMERICAN INDIANS, Lewis Spence. Myths and legends of the Algonquins, Iroquois, Pawnees and Sioux with comprehensive historical and ethnological commentary. 36 illustrations. 5⅜ × 8½.
25967-6 Pa. $8.95

GREAT DINOSAUR HUNTERS AND THEIR DISCOVERIES, Edwin H. Colbert. Fascinating, lavishly illustrated chronicle of dinosaur research, 1820s to 1960. Achievements of Cope, Marsh, Brown, Buckland, Mantell, Huxley, many others. 384pp. 5¼ × 8¼. 24701-5 Pa. $7.95

THE TASTEMAKERS, Russell Lynes. Informal, illustrated social history of American taste 1850s–1950s. First popularized categories Highbrow, Lowbrow, Middlebrow. 129 illustrations. New (1979) afterword. 384pp. 6 × 9.
23993-4 Pa. $8.95

DOUBLE CROSS PURPOSES, Ronald A. Knox. A treasure hunt in the Scottish Highlands, an old map, unidentified corpse, surprise discoveries keep reader guessing in this cleverly intricate tale of financial skullduggery. 2 black-and-white maps. 320pp. 5⅜ × 8½. (Available in U.S. only) 25032-6 Pa. $6.95

AUTHENTIC VICTORIAN DECORATION AND ORNAMENTATION IN FULL COLOR: 46 Plates from "Studies in Design," Christopher Dresser. Superb full-color lithographs reproduced from rare original portfolio of a major Victorian designer. 48pp. 9¼ × 12¼. 25083-0 Pa. $7.95

PRIMITIVE ART, Franz Boas. Remains the best text ever prepared on subject, thoroughly discussing Indian, African, Asian, Australian, and, especially, Northern American primitive art. Over 950 illustrations show ceramics, masks, totem poles, weapons, textiles, paintings, much more. 376pp. 5⅜ × 8. 20025-6 Pa. $7.95

SIDELIGHTS ON RELATIVITY, Albert Einstein. Unabridged republication of two lectures delivered by the great physicist in 1920–21. *Ether and Relativity* and *Geometry and Experience*. Elegant ideas in nonmathematical form, accessible to intelligent layman. vi + 56pp. 5⅜ × 8½. 24511-X Pa. $3.95

THE WIT AND HUMOR OF OSCAR WILDE, edited by Alvin Redman. More than 1,000 ripostes, paradoxes, wisecracks: Work is the curse of the drinking classes, I can resist everything except temptation, etc. 258pp. 5⅜ × 8½. 20602-5 Pa. $4.95

ADVENTURES WITH A MICROSCOPE, Richard Headstrom. 59 adventures with clothing fibers, protozoa, ferns and lichens, roots and leaves, much more. 142 illustrations. 232pp. 5⅜ × 8½. 23471-1 Pa. $4.95

PLANTS OF THE BIBLE, Harold N. Moldenke and Alma L. Moldenke. Standard reference to all 230 plants mentioned in Scriptures. Latin name, biblical reference, uses, modern identity, much more. Unsurpassed encyclopedic resource for scholars, botanists, nature lovers, students of Bible. Bibliography. Indexes. 123 black-and-white illustrations. 384pp. 6 × 9. 25069-5 Pa. $8.95

FAMOUS AMERICAN WOMEN: A Biographical Dictionary from Colonial Times to the Present, Robert McHenry, ed. From Pocahontas to Rosa Parks, 1,035 distinguished American women documented in separate biographical entries. Accurate, up-to-date data, numerous categories, spans 400 years. Indices. 493pp. 6½ × 9¼. 24523-3 Pa. $10.95

THE FABULOUS INTERIORS OF THE GREAT OCEAN LINERS IN HISTORIC PHOTOGRAPHS, William H. Miller, Jr. Some 200 superb photographs capture exquisite interiors of world's great "floating palaces"—1890s to 1980s: *Titanic, Ile de France, Queen Elizabeth, United States, Europa,* more. Approx. 200 black-and-white photographs. Captions. Text. Introduction. 160pp. 8⅞ × 11¾. 24756-2 Pa. $9.95

THE GREAT LUXURY LINERS, 1927–1954: A Photographic Record, William H. Miller, Jr. Nostalgic tribute to heyday of ocean liners. 186 photos of *Ile de France, Normandie, Leviathan, Queen Elizabeth, United States,* many others. Interior and exterior views. Introduction. Captions. 160pp. 9 × 12. 24056-8 Pa. $12.95

A NATURAL HISTORY OF THE DUCKS, John Charles Phillips. Great landmark of ornithology offers complete detailed coverage of nearly 200 species and subspecies of ducks: gadwall, sheldrake, merganser, pintail, many more. 74 full-color plates, 102 black-and-white. Bibliography. Total of 1,920pp. 8⅝ × 11¼. 25141-1, 25142-X Cloth., Two-vol. set $100.00

THE SEAWEED HANDBOOK: An Illustrated Guide to Seaweeds from North Carolina to Canada, Thomas F. Lee. Concise reference covers 78 species. Scientific and common names, habitat, distribution, more. Finding keys for easy identification. 224pp. 5⅜ × 8½. 25215-9 Pa. $6.95

THE TEN BOOKS OF ARCHITECTURE: The 1755 Leoni Edition, Leon Battista Alberti. Rare classic helped introduce the glories of ancient architecture to the Renaissance. 68 black-and-white plates. 336pp. 8⅜ × 11¼. 25239-6 Pa. $14.95

MISS MACKENZIE, Anthony Trollope. Minor masterpieces by Victorian master unmasks many truths about life in 19th-century England. First inexpensive edition in years. 392pp. 5⅜ × 8½. 25201-9 Pa. $8.95

THE RIME OF THE ANCIENT MARINER, Gustave Doré, Samuel Taylor Coleridge. Dramatic engravings considered by many to be his greatest work. The terrifying space of the open sea, the storms and whirlpools of an unknown ocean, the ice of Antarctica, more—all rendered in a powerful, chilling manner. Full text. 38 plates. 77pp. 9¼ × 12. 22305-1 Pa. $4.95

THE EXPEDITIONS OF ZEBULON MONTGOMERY PIKE, Zebulon Montgomery Pike. Fascinating firsthand accounts (1805–6) of exploration of Mississippi River, Indian wars, capture by Spanish dragoons, much more. 1,088pp. 5⅜ × 8½. 25254-X, 25255-8 Pa., Two-vol. set $25.90

CATALOG OF DOVER BOOKS

A CONCISE HISTORY OF PHOTOGRAPHY: Third Revised Edition, Helmut Gernsheim. Best one-volume history—camera obscura, photochemistry, daguerreotypes, evolution of cameras, film, more. Also artistic aspects—landscape, portraits, fine art, etc. 281 black-and-white photographs. 26 in color. 176pp. 8⅜ × 11¼.
25128-4 Pa. $14.95

THE DORÉ BIBLE ILLUSTRATIONS, Gustave Doré. 241 detailed plates from the Bible: the Creation scenes, Adam and Eve, Flood, Babylon, battle sequences, life of Jesus, etc. Each plate is accompanied by the verses from the King James version of the Bible. 241pp. 9 × 12. 23004-X Pa. $9.95

WANDERINGS IN WEST AFRICA, Richard F. Burton. Great Victorian scholar/adventurer's invaluable descriptions of African tribal rituals, fetishism, culture, art, much more. Fascinating 19th-century account. 624pp. 5⅜ × 8½. 26890-X Pa. $12.95

HISTORIC HOMES OF THE AMERICAN PRESIDENTS, Second Revised Edition, Irvin Haas. Guide to homes occupied by every president from Washington to Bush. Visiting hours, travel routes, more. 175 photos. 160pp. 8¼ × 11.
26751-2 Pa. $9.95

THE HISTORY OF THE LEWIS AND CLARK EXPEDITION, Meriwether Lewis and William Clark, edited by Elliott Coues. Classic edition of Lewis and Clark's day-by-day journals that later became the basis for U.S. claims to Oregon and the West. Accurate and invaluable geographical, botanical, biological, meteorological and anthropological material. Total of 1,508pp. 5⅜ × 8½.
21268-8, 21269-6, 21270-X Pa., Three-vol. set $29.85

LANGUAGE, TRUTH AND LOGIC, Alfred J. Ayer. Famous, clear introduction to Vienna, Cambridge schools of Logical Positivism. Role of philosophy, elimination of metaphysics, nature of analysis, etc. 160pp. 5⅜ × 8½. (Available in U.S. and Canada only) 20010-8 Pa. $3.95

MATHEMATICS FOR THE NONMATHEMATICIAN, Morris Kline. Detailed, college-level treatment of mathematics in cultural and historical context, with numerous exercises. For liberal arts students. Preface. Recommended Reading Lists. Tables. Index. Numerous black-and-white figures. xvi + 641pp. 5⅜ × 8½.
24823-2 Pa. $11.95

HANDBOOK OF PICTORIAL SYMBOLS, Rudolph Modley. 3,250 signs and symbols, many systems in full; official or heavy commercial use. Arranged by subject. Most in Pictorial Archive series. 143pp. 8⅛ × 11. 23357-X Pa. $7.95

INCIDENTS OF TRAVEL IN YUCATAN, John L. Stephens. Classic (1843) exploration of jungles of Yucatan, looking for evidences of Maya civilization. Travel adventures, Mexican and Indian culture, etc. Total of 669pp. 5⅜ × 8½.
20926-1, 20927-X Pa., Two-vol. set $13.90

CATALOG OF DOVER BOOKS

DEGAS: An Intimate Portrait, Ambroise Vollard. Charming, anecdotal memoir by famous art dealer of one of the greatest 19th-century French painters. 14 black-and-white illustrations. Introduction by Harold L. Van Doren. 96pp. 5⅜ × 8½.
25131-4 Pa. $4.95

PERSONAL NARRATIVE OF A PILGRIMAGE TO AL-MADINAH AND MECCAH, Richard F. Burton. Great travel classic by remarkably colorful personality. Burton, disguised as a Moroccan, visited sacred shrines of Islam, narrowly escaping death. 47 illustrations. 959pp. 5⅜ × 8½.
21217-3, 21218-1 Pa., Two-vol. set $19.90

PHRASE AND WORD ORIGINS, A. H. Holt. Entertaining, reliable, modern study of more than 1,200 colorful words, phrases, origins and histories. Much unexpected information. 254pp. 5⅜ × 8½.
20758-7 Pa. $5.95

THE RED THUMB MARK, R. Austin Freeman. In this first Dr. Thorndyke case, the great scientific detective draws fascinating conclusions from the nature of a single fingerprint. Exciting story, authentic science. 320pp. 5⅜ × 8½. (Available in U.S. only)
25210-8 Pa. $6.95

AN EGYPTIAN HIEROGLYPHIC DICTIONARY, E. A. Wallis Budge. Monumental work containing about 25,000 words or terms that occur in texts ranging from 3000 B.C. to 600 A.D. Each entry consists of a transliteration of the word, the word in hieroglyphs, and the meaning in English. 1,314pp. 6⅝ × 10.
23615-3, 23616-1 Pa., Two-vol. set $35.90

THE COMPLEAT STRATEGYST: Being a Primer on the Theory of Games of Strategy, J. D. Williams. Highly entertaining classic describes, with many illustrated examples, how to select best strategies in conflict situations. Prefaces. Appendices. xvi + 268pp. 5⅜ × 8½.
25101-2 Pa. $6.95

THE ROAD TO OZ, L. Frank Baum. Dorothy meets the Shaggy Man, little Button-Bright and the Rainbow's beautiful daughter in this delightful trip to the magical Land of Oz. 272pp. 5⅜ × 8.
25208-6 Pa. $5.95

POINT AND LINE TO PLANE, Wassily Kandinsky. Seminal exposition of role of point, line, other elements in nonobjective painting. Essential to understanding 20th-century art. 127 illustrations. 192pp. 6½ × 9¼.
23808-3 Pa. $5.95

LADY ANNA, Anthony Trollope. Moving chronicle of Countess Lovel's bitter struggle to win for herself and daughter Anna their rightful rank and fortune—perhaps at cost of sanity itself. 384pp. 5⅜ × 8½.
24669-8 Pa. $8.95

EGYPTIAN MAGIC, E. A. Wallis Budge. Sums up all that is known about magic in Ancient Egypt: the role of magic in controlling the gods, powerful amulets that warded off evil spirits, scarabs of immortality, use of wax images, formulas and spells, the secret name, much more. 253pp. 5⅜ × 8½.
22681-6 Pa. $4.95

THE DANCE OF SIVA, Ananda Coomaraswamy. Preeminent authority unfolds the vast metaphysic of India: the revelation of her art, conception of the universe, social organization, etc. 27 reproductions of art masterpieces. 192pp. 5⅜ × 8½.
24817-8 Pa. $6.95

CHRISTMAS CUSTOMS AND TRADITIONS, Clement A. Miles. Origin, evolution, significance of religious, secular practices. Caroling, gifts, yule logs, much more. Full, scholarly yet fascinating; non-sectarian. 400pp. 5⅜ × 8½.
23354-5 Pa. $7.95

THE HUMAN FIGURE IN MOTION, Eadweard Muybridge. More than 4,500 stopped-action photos, in action series, showing undraped men, women, children jumping, lying down, throwing, sitting, wrestling, carrying, etc. 390pp. 7⅞ × 10⅝.
20204-6 Cloth. $24.95

THE MAN WHO WAS THURSDAY, Gilbert Keith Chesterton. Witty, fast-paced novel about a club of anarchists in turn-of-the-century London. Brilliant social, religious, philosophical speculations. 128pp. 5⅜ × 8½.
25121-7 Pa. $3.95

A CÉZANNE SKETCHBOOK: Figures, Portraits, Landscapes and Still Lifes, Paul Cézanne. Great artist experiments with tonal effects, light, mass, other qualities in over 100 drawings. A revealing view of developing master painter, precursor of Cubism. 102 black-and-white illustrations. 144pp. 8¾ × 6⅝.
24790-2 Pa. $6.95

AN ENCYCLOPEDIA OF BATTLES: Accounts of Over 1,560 Battles from 1479 B.C. to the Present, David Eggenberger. Presents essential details of every major battle in recorded history, from the first battle of Megiddo in 1479 B.C. to Grenada in 1984. List of Battle Maps. New Appendix covering the years 1967–1984. Index. 99 illustrations. 544pp. 6½ × 9¼.
24913-1 Pa. $14.95

AN ETYMOLOGICAL DICTIONARY OF MODERN ENGLISH, Ernest Weekley. Richest, fullest work, by foremost British lexicographer. Detailed word histories. Inexhaustible. Total of 856pp. 6½ × 9¼.
21873-2, 21874-0 Pa., Two-vol. set $19.90

WEBSTER'S AMERICAN MILITARY BIOGRAPHIES, edited by Robert McHenry. Over 1,000 figures who shaped 3 centuries of American military history. Detailed biographies of Nathan Hale, Douglas MacArthur, Mary Hallaren, others. Chronologies of engagements, more. Introduction. Addenda. 1,033 entries in alphabetical order. xi + 548pp. 6½ × 9¼. (Available in U.S. only)
24758-9 Pa. $13.95

LIFE IN ANCIENT EGYPT, Adolf Erman. Detailed older account, with much not in more recent books: domestic life, religion, magic, medicine, commerce, and whatever else needed for complete picture. Many illustrations. 597pp. 5⅜ × 8½.
22632-8 Pa. $9.95

HISTORIC COSTUME IN PICTURES, Braun & Schneider. Over 1,450 costumed figures shown, covering a wide variety of peoples: kings, emperors, nobles, priests, servants, soldiers, scholars, townsfolk, peasants, merchants, courtiers, cavaliers, and more. 256pp. 8⅜ × 11¼.
23150-X Pa. $9.95

THE NOTEBOOKS OF LEONARDO DA VINCI, edited by J. P. Richter. Extracts from manuscripts reveal great genius; on painting, sculpture, anatomy, sciences, geography, etc. Both Italian and English. 186 ms. pages reproduced, plus 500 additional drawings, including studies for *Last Supper, Sforza* monument, etc. 860pp. 7⅞ × 10¾. (Available in U.S. only) 22572-0, 22573-9 Pa., Two-vol. set $35.90

THE ART NOUVEAU STYLE BOOK OF ALPHONSE MUCHA: All 72 Plates from "Documents Decoratifs" in Original Color, Alphonse Mucha. Rare copyright-free design portfolio by high priest of Art Nouveau. Jewelry, wallpaper, stained glass, furniture, figure studies, plant and animal motifs, etc. Only complete one-volume edition. 80pp. 9⅜ × 12¼. 24044-4 Pa. $9.95

ANIMALS: 1,419 COPYRIGHT-FREE ILLUSTRATIONS OF MAMMALS, BIRDS, FISH, INSECTS, ETC., edited by Jim Harter. Clear wood engravings present, in extremely lifelike poses, over 1,000 species of animals. One of the most extensive pictorial sourcebooks of its kind. Captions. Index. 284pp. 9 × 12. 23766-4 Pa. $9.95

OBELISTS FLY HIGH, C. Daly King. Masterpiece of American detective fiction, long out of print, involves murder on a 1935 transcontinental flight—"a very thrilling story"—NY Times. Unabridged and unaltered republication of the edition published by William Collins Sons & Co. Ltd., London, 1935. 288pp. 5⅜ × 8½. (Available in U.S. only) 25036-9 Pa. $5.95

VICTORIAN AND EDWARDIAN FASHION: A Photographic Survey, Alison Gernsheim. First fashion history completely illustrated by contemporary photographs. Full text plus 235 photos, 1840-1914, in which many celebrities appear. 240pp. 6½ × 9¼. 24205-6 Pa. $8.95

THE ART OF THE FRENCH ILLUSTRATED BOOK, 1700-1914, Gordon N. Ray. Over 630 superb book illustrations by Fragonard, Delacroix, Daumier, Doré, Grandville, Manet, Mucha, Steinlen, Toulouse-Lautrec and many others. Preface. Introduction. 633 halftones. Indices of artists, authors & titles, binders and provenances. Appendices. Bibliography. 608pp. 8⅜ × 11¼. 25086-5 Pa. $24.95

THE WONDERFUL WIZARD OF OZ, L. Frank Baum. Facsimile in full color of America's finest children's classic. 143 illustrations by W. W. Denslow. 267pp. 5⅜ × 8½. 20691-2 Pa. $7.95

FOLLOWING THE EQUATOR: A Journey Around the World, Mark Twain. Great writer's 1897 account of circumnavigating the globe by steamship. Ironic humor, keen observations, vivid and fascinating descriptions of exotic places. 197 illustrations. 720pp. 5⅜ × 8½. 26113-1 Pa. $15.95

THE FRIENDLY STARS, Martha Evans Martin & Donald Howard Menzel. Classic text marshalls the stars together in an engaging, non-technical survey, presenting them as sources of beauty in night sky. 23 illustrations. Foreword. 2 star charts. Index. 147pp. 5⅜ × 8½. 21099-5 Pa. $3.95

FADS AND FALLACIES IN THE NAME OF SCIENCE, Martin Gardner. Fair, witty appraisal of cranks, quacks, and quackeries of science and pseudoscience: hollow earth, Velikovsky, orgone energy, Dianetics, flying saucers, Bridey Murphy, food and medical fads, etc. Revised, expanded In the Name of Science. "A very able and even-tempered presentation."—The New Yorker. 363pp. 5⅜ × 8. 20394-8 Pa. $6.95

ANCIENT EGYPT: ITS CULTURE AND HISTORY, J. E Manchip White. From pre-dynastics through Ptolemies: society, history, political structure, religion, daily life, literature, cultural heritage. 48 plates. 217pp. 5⅜ × 8½. 22548-8 Pa. $5.95

CATALOG OF DOVER BOOKS

SIR HARRY HOTSPUR OF HUMBLETHWAITE, Anthony Trollope. Incisive, unconventional psychological study of a conflict between a wealthy baronet, his idealistic daughter, and their scapegrace cousin. The 1870 novel in its first inexpensive edition in years. 250pp. 5⅜ × 8½.   24953-0 Pa. $6.95

LASERS AND HOLOGRAPHY, Winston E. Kock. Sound introduction to burgeoning field, expanded (1981) for second edition. Wave patterns, coherence, lasers, diffraction, zone plates, properties of holograms, recent advances. 84 illustrations. 160pp. 5⅜ × 8¼. (Except in United Kingdom)   24041-X Pa. $3.95

INTRODUCTION TO ARTIFICIAL INTELLIGENCE: Second, Enlarged Edition, Philip C. Jackson, Jr. Comprehensive survey of artificial intelligence—the study of how machines (computers) can be made to act intelligently. Includes introductory and advanced material. Extensive notes updating the main text. 132 black-and-white illustrations. 512pp. 5⅜ × 8½.   24864-X Pa. $10.95

HISTORY OF INDIAN AND INDONESIAN ART, Ananda K. Coomaraswamy. Over 400 illustrations illuminate classic study of Indian art from earliest Harappa finds to early 20th century. Provides philosophical, religious and social insights. 304pp. 6⅝ × 9⅜.   25005-9 Pa. $11.95

THE GOLEM, Gustav Meyrink. Most famous supernatural novel in modern European literature, set in Ghetto of Old Prague around 1890. Compelling story of mystical experiences, strange transformations, profound terror. 13 black-and-white illustrations. 224pp. 5⅜ × 8½. (Available in U.S. only)   25025-3 Pa. $6.95

PICTORIAL ENCYCLOPEDIA OF HISTORIC ARCHITECTURAL PLANS, DETAILS AND ELEMENTS: With 1,880 Line Drawings of Arches, Domes, Doorways, Facades, Gables, Windows, etc., John Theodore Haneman. Sourcebook of inspiration for architects, designers, others. Bibliography. Captions. 141pp. 9 × 12.
24605-1 Pa. $8.95

BENCHLEY LOST AND FOUND, Robert Benchley. Finest humor from early 30s, about pet peeves, child psychologists, post office and others. Mostly unavailable elsewhere. 73 illustrations by Peter Arno and others. 183pp. 5⅜ × 8½.
22410-4 Pa. $4.95

ERTÉ GRAPHICS, Erté. Collection of striking color graphics: *Seasons, Alphabet, Numerals, Aces* and *Precious Stones.* 50 plates, including 4 on covers. 48pp. 9⅜ × 12¼.
23580-7 Pa. $7.95

THE JOURNAL OF HENRY D. THOREAU, edited by Bradford Torrey, F. H. Allen. Complete reprinting of 14 volumes; 1837–61, over two million words; the sourcebooks for *Walden,* etc. Definitive. All original sketches, plus 75 photographs. 1,804pp. 8½ × 12¼.   20312-3, 20313-1 Cloth., Two-vol. set $130.00

CASTLES: Their Construction and History, Sidney Toy. Traces castle development from ancient roots. Nearly 200 photographs and drawings illustrate moats, keeps, baileys, many other features. Caernarvon, Dover Castles, Hadrian's Wall, Tower of London, dozens more. 256pp. 5⅜ × 8¼.   24898-4 Pa. $7.95

AMERICAN CLIPPER SHIPS: 1833–1858, Octavius T. Howe & Frederick C. Matthews. Fully-illustrated, encyclopedic review of 352 clipper ships from the period of America's greatest maritime supremacy. Introduction. 109 halftones. 5 black-and-white line illustrations. Index. Total of 928pp. 5⅜ × 8½.
25115-2, 25116-0 Pa., Two-vol. set $17.90

TOWARDS A NEW ARCHITECTURE, Le Corbusier. Pioneering manifesto by great architect, near legendary founder of "International School." Technical and aesthetic theories, views on industry, economics, relation of form to function, "mass-production spirit," much more. Profusely illustrated. Unabridged translation of 13th French edition. Introduction by Frederick Etchells. 320pp. 6⅛ × 9¼. (Available in U.S. only)
25023-7 Pa. $8.95

THE BOOK OF KELLS, edited by Blanche Cirker. Inexpensive collection of 32 full-color, full-page plates from the greatest illuminated manuscript of the Middle Ages, painstakingly reproduced from rare facsimile edition. Publisher's Note. Captions. 32pp. 9⅜ × 12¼. (Available in U.S. only)
24345-1 Pa. $5.95

BEST SCIENCE FICTION STORIES OF H. G. WELLS, H. G. Wells. Full novel *The Invisible Man*, plus 17 short stories: "The Crystal Egg," "Aepyornis Island," "The Strange Orchid," etc. 303pp. 5⅜ × 8½. (Available in U.S. only)
21531-8 Pa. $6.95

AMERICAN SAILING SHIPS: Their Plans and History, Charles G. Davis. Photos, construction details of schooners, frigates, clippers, other sailcraft of 18th to early 20th centuries—plus entertaining discourse on design, rigging, nautical lore, much more. 137 black-and-white illustrations. 240pp. 6⅛ × 9¼.
24658-2 Pa. $6.95

ENTERTAINING MATHEMATICAL PUZZLES, Martin Gardner. Selection of author's favorite conundrums involving arithmetic, money, speed, etc., with lively commentary. Complete solutions. 112pp. 5⅜ × 8½.
25211-6 Pa. $3.50

THE WILL TO BELIEVE, HUMAN IMMORTALITY, William James. Two books bound together. Effect of irrational on logical, and arguments for human immortality. 402pp. 5⅜ × 8½.
20291-7 Pa. $8.95

THE HAUNTED MONASTERY and THE CHINESE MAZE MURDERS, Robert Van Gulik. 2 full novels by Van Gulik continue adventures of Judge Dee and his companions. An evil Taoist monastery, seemingly supernatural events; overgrown topiary maze that hides strange crimes. Set in 7th-century China. 27 illustrations. 328pp. 5⅜ × 8½.
23502-5 Pa. $6.95

CELEBRATED CASES OF JUDGE DEE (DEE GOONG AN), translated by Robert Van Gulik. Authentic 18th-century Chinese detective novel; Dee and associates solve three interlocked cases. Led to Van Gulik's own stories with same characters. Extensive introduction. 9 illustrations. 237pp. 5⅜ × 8½.
23337-5 Pa. $5.95

Prices subject to change without notice.

Available at your book dealer or write for free catalog to Dept. GI, Dover Publications, Inc., 31 East 2nd St., Mineola, N.Y. 11501. Dover publishes more than 175 books each year on science, elementary and advanced mathematics, biology, music, art, literary history, social sciences and other areas.